L.S. Pontrjagin   Learning Higher Mathematics

Lev S. Pontrjagin

Learning Higher Mathematics

Part I
The Method of Coordinates

Part II
Analysis of the Infinitely Small

Translated from the Russian
by Edwin Hewitt

With 68 Figures

Springer-Verlag
Berlin Heidelberg New York Tokyo
1984

Lev Semenovič Pontrjagin
Member of the Academy of Sciences of the USSR
Professor of Mathematics, Steklov Mathematical Institute
ul. Vavilova 42, Moscow 117333, USSR

Edwin Hewitt
Professor of Mathematics, University of Washington
Seattle, WA 98195, USA

Titles of the Russian original editions:
Metod koordinat
Analiz beskonechno malykh
Publisher Nauka, Moscow 1977, 1980

This volume is part of the *Springer Series in Soviet Mathematics*
Advisers: L.D. Faddeev (Leningrad), R.V. Gamkrelidze (Moscow)

AMS Subject Classification: 15-01, 26-01, 30-01, 40-01

ISBN 3-540-12351-2 Springer-Verlag Berlin Heidelberg New York Tokyo
ISBN 0-387-12351-2 Springer-Verlag New York Heidelberg Berlin Tokyo

Library of Congress Cataloging in Publication Data
Pontrjagin, L.S. (Lev Semenovich), 1908
Learning higher mathematics. (Springer series in Soviet mathematics)
Translation of: Metod koordinat and Analiz beskonechno malykh.
Includes index. Contents: pt. 1. The method of coordinates – pt. 2. Analysis of the infinitely small.
1. Coordinates. 2. Mathematical analysis. I. Pontrjagin, L.S. (Lev Semenovich), 1908.
Analiz beskonechno malykh. English. 1984. II. Title. III. Series.
QA556. P65713   1984   515   83-14823
ISBN 0-387-12351-2 (U.S.)

© by Springer-Verlag Berlin Heidelberg 1984
Printed in Germany
Typesetting, printing and binding: Universitätsdruckerei H. Stürtz AG, D-8700 Würzburg
2141/3140-543210

Translator's Preface

Lev Semenovič Pontrjagin (1908) is one of the outstanding figures in 20th century mathematics. In a long career he has made fundamental contributions to many branches of mathematics, both pure and applied. He has received every honor that a grateful government can bestow. Though in no way constrained to do so, he has through the years taught mathematics courses at Moscow State University. In the year 1975 he set himself the task of writing a series of books on secondary school and beginning university mathematics. In his own words, "I wished to set forth the foundations of higher mathematics in a form that would have been accessible to myself as a lad, but making use of all my experience as a scientist and a teacher, accumulated over many years." The present volume is a translation of the first two out of four moderately sized volumes on this theme planned by Professor Pontrjagin.

The book begins at the beginning of modern mathematics, analytic geometry in the plane and 3-dimensional space. Refinements about limits and the nature of real numbers come only later. Many concrete examples are given; these may take the place of formal exercises, which the book does not provide. The book continues with careful treatment of differentiation and integration, of limits, of expansions of elementary functions in power series. The final sections deal with analytic functions of a complex variable, ending with a proof of a famous theorem about the behavior of an analytic function near an essential singularity.

The book bears throughout the stamp of the careful thought and constant pursuit of clarity that characterize all of Professor Pontrjagin's scientific writing.

The book cannot be said to be "easy reading." By its very nature mathematics is hard. But the reader who perseveres in studying this book will reap a rich reward in both knowledge and satisfaction.

The translator's thanks are due to Mary Keeler, who rendered vital assistance in redrawing a figure from the original.

Seattle, Washington, July 1983   Edwin Hewitt

Table of Contents

Part I
The Method of Coordinates

Part II
Analysis of the Infinitely Small

Part I
The Method of Coordinates

Introduction to Part I

At the present day, scientists in many different fields know about rectangular Cartesian coordinates in the plane. This is because such coordinates give a visual, geometric representation by graphs of how one variable varies with another. Thus a physician may plot a graph of the temperature of a patient during the course of an illness. An economist may plot the growth of production of industrial goods. There are a host of examples. The name *Cartesian Coordinates* may lead to the false impression that Descartes invented Cartesian coordinates. In point of fact, rectangular coordinates have been used at least since the beginning of the Christian era. Descartes perfected rectangular coordinates in a very important point, namely, with his rule for choice of signs (see below). But most important of all, he introduced analytic geometry, thus uniting geometry with algebra. One should also point out that Descartes' discovery was made simultaneously by another French mathematician, Fermat, who is known to a wide public by his "last theorem", which remains unproved to this day. By the 17th century, when Descartes and Fermat lived, the development of mathematics had prepared the soil for the discovery of analytic geometry, which is to say, the synthesis of algebra and geometry. The last blow was struck by Descartes and Fermat. Curiously, nearly everyone knows only about Descartes. This is evidently the fate of all or almost all great scientific discoveries. They are prepared over centuries, and the final step, made by some scientist, is associated with his name. Then the whole discovery bears his name. I will give below a short and very incomplete sketch of the development of mathematics up to the 17th century in order to illustrate my thesis by the example of analytic geometry.

To construct a system of coordinates we make a diagram of two lines in the plane. These are the axes of coordinates. One of them is horizontal and is called the axis of abscissas. The other is vertical and is called the axis of ordinates. The intersection o of the axes is called the origin of coordinates or simply origin. Let r be any point in the plane containing our diagram. From r we drop a perpendicular rp reaching to the axis of abscissas and a perpendicular rq reaching to the axis of ordinates. With this construction we can associate two nonnegative numbers with the point r: the length of the line segment op and the length of the line segment oq. Suppose that r' is the point symmetric with r with respect to the axis of ordinates. The numbers associated with r and r' are plainly equal. The same is true for the point symmetric to r with respect to the axis of abscissas. This observation explains Descartes' rule for choice of signs. If the point r lies to the right of the axis of ordinates, then the number oq is to be taken with a plus sign. If r lies to the left of the axis of ordinates, the number oq is to be taken with a minus sign. We assign a plus or a minus sign to the number op analogously, as r lies above or below the axis of abscissas. The number oq with a plus or minus sign as described is ordinarily denoted by x. The number op with plus or minus sign as described is ordinarily denoted by y. The numbers x and y are called *the Cartesian coordinates of the point* r. If the variable quantities x and y are connected by some relation, for example an algebraic equation, then the set of all points r whose coordinates satisfy this equation form a line or curve in the plane. In particular, one constructs a graph in just this way, if y is determined by x.

The construction described in the last paragraph is firmly associated with the name of Descartes (1596–1650). However, in one or another less complete form it was used by mathematicians long before his time. The ancient Alexandrian mathematician Apollonius (3 or 2 centuries B.C.) used what amount to rectangular coordinates. Using them, he defined the parabola, the hyperbola, and the ellipse, which were studied in great detail and were well known in his era. Apollonius found equations for these curves:

$$y^2 = px \text{ (parabola)};$$

$$y^2 = px + \frac{p}{a}x^2 \text{ (hyperbola)};$$

$$y^2 = px - \frac{p}{a}x^2 \text{ (ellipse)}.$$

Here p and a are positive constants.

Apollonius naturally did not write the equations in the above algebraic form, since in his time algebraic symbolism was unknown. Instead he wrote his equations using geometric concepts: y^2 is the area of a square of side y; px is the area of a rectangle with sides p and x; and so on. The names of the curves are connected with these equations. In Greek, parabola means equality: the square y^2 has area equal to the rectangle px. In Greek, hyperbola means excess: the area of the square y^2 exceeds the area of the rectangle px. In Greek, ellipse means deficit: the area of the square y^2 is less than the area of the rectangle px.

The French mathematician Nicole Oresme (c. 1323–1382) used rectangular coordinates to construct the graph showing the dependence of a variable y on a variable x. He did not use the terms abscissa and ordinate, common today, but instead the Latin words meaning longitude and latitude, respectively. Nevertheless, the ideas of Oresme received little distribution, since in his time the notion of functional dependence (that is, how a variable y can depend upon a variable x) was entirely too obscure.

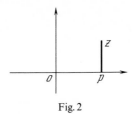

Fig. 2

lies to the right of o if x is positive and to the left of o if x is negative. We now erect a perpendicular pz to the axis of abscissas whose length is $|y|$. The point z lies above the axis of abscissas if y is positive and below the axis of abscissas if y is negative. The end z of this perpendicular is the point z whose coordinates are x and y. Thus formula (1) holds for this point z, and so (1) establishes a one-to-one correspondence between points of the plane P and pairs of numbers (x, y).

Suppose that a point z moves along the axis of abscissas from left to right. Then its abscissa increases. For this reason one says that the axis of abscissas is oriented from left to right. We indicate this pictorially by sketching an arrowhead on the axis of abscissas, aimed to the right (again see Fig. 2). In the same sense, the axis of ordinates is oriented from below to above. We indicate this pictorially with an arrowhead on the axis of ordinates that is aimed upward. Points on the axis of abscissas lying to the right of 0 have positive abscissas. Thus one says that they comprise *the positive half-axis of abscissas*. Similarly, the points on the axis of abscissas to the left of 0 comprise *the negative half-axis of abscissas*. The positive and negative half-axes of ordinates are defined similarly.

The first problem that we will solve by the method of coordinates is to compute the length of a line segment. This is equivalent to finding the distance between two points in terms of their coordinates.

Suppose that

$$z_1 = (x_1, y_1) \quad \text{and} \quad z_2 = (x_2, y_2) \tag{2}$$

are any two points in the plane P. Let $l(z_1, z_2)$ denote the *distance* between the points z_1 and z_2. We then have the formula[1]

$$l(z_1, z_2) = +\sqrt{(x_2 - x_1)^2 + (y_2 - y_1)^2}. \tag{3}$$

To prove formula (3), we proceed as in Fig. 3. Draw a horizontal line through z_1 and a vertical line through z_2. Let r be the point where these two lines intersect. The points z_1, r, z_2 are vertices of a right triangle with

---

[1] Our convention regarding the sign of square roots is the following. The sign "$+$" indicates the positive square root, the sign "$-$" the negative square root. If neither sign appears before the square root symbol, then either the positive or the negative square root may be chosen.

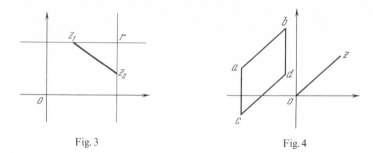

Fig. 3   Fig. 4

legs $z_1 r$ and $r z_2$ and hypotenuse $z_1 z_2$. It is obvious that the length of the
line segment $z_1 r$ is equal to $|x_2 - x_1|$ and that the length of the line segment
$r z_2$ is equal to $|y_2 - y_1|$. Accordingly (3) follows from the theorem of
Pythagoras, applied to the right triangle $z_1 r z_2$.

It is now appropriate to turn to a consideration of vectors, which enable
us to write many formulas much more simply than with the use of coordinates.

The reader may now turn to Fig. 4. Let a and b be any points in the
plane, as sketched in Fig. 4. Consider the line segment ab, considered as
beginning at a and ending at b. We refer to this directed line segment as *the
vector ab*. The point a is called *the initial point* and the point b *the terminal
point*, of the vector ab. One also says that the vector ab is *affixed to the
point a*. Let c and d be any other pair of points in the plane P. We say that
the vector ab *is equal to the vector cd* if the line segments ab and cd are: 1)
of the same length; 2) parallel; 3) of the same orientation. When the line
segments ab and cd do not lie on a single line, we can restate these
conditions in a simple geometric form. Draw the line segments ac and bd in
addition to ab and cd (again see Fig. 4). Conditions 1)–3) are equivalent to
the single requirement that $acdb$ be a parallelogram.

It is clear that for every vector ab there is a vector equal to ab whose
initial point is the origin o. Denote its terminal point by z, so that ab and
oz are equal vectors. We agree to write vectors of the form oz by the single
letter z. This expresses a one-to-one correspondence between points in the
plane and vectors with initial point o: with the vector oz we associate its
terminal point z. We shall encounter no confusion in writing both the point
and the vector by the same symbol.

We now define the sum of two vectors. Let z_1 and z_2 be two vectors in
the plane P. We keep in mind that z_1 and z_2 are the terminal points of two
vectors with initial point o, as sketched in Fig. 5. Now construct the vector
$z_1 z_3$, defining z_3 by the requirement that $z_1 z_3$ be equal to the vector z_2.
The vector oz_3, or simply z_3 under our convention, is defined to be the sum
of the vectors z_1 and z_2:

$$z_3 = z_1 + z_2.$$

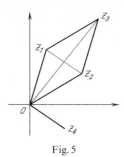

Fig. 5

In our construction of the sum, z_1 and z_2 play different rôles. Hence we have to prove that

$$z_1 + z_2 = z_2 + z_1.$$

To do this, we construct the vector $z_2 z_3$. Since the vectors $o z_2$ and $z_1 z_3$ are equal, the figure $0 z_1 z_3 z_2$ is a parallelogram. Hence the vectors $o z_1$ and $z_2 z_3$ are equal. By definition, therefore, z_3 is the sum $z_2 + z_1$. That is to say, the sum of two vectors is independent of the order of summation.

  The difference $z_1 - z_2$ of the vectors z_1 and z_2 is defined as the vector $z_1 z_2$ (again see Fig. 5). Let us write it as z_4. This means simply that the vector $o z_4$ is equal to the vector $z_1 z_2$. From the definition of sum we find that

$$z_1 + z_4 = z_2.$$

That is, the difference $z_2 - z_1$ is so defined that

$$(z_2 - z_1) + z_1 = z_2. \tag{4}$$

Formula (4) indicates that our definition of the difference of two vectors was correct. Remember that in defining $z_2 - z_1$, we take z_1 as the initial point and z_2 as the terminal point.

  There is exactly one vector of zero length: its initial and terminal points coincide, and can be taken as o. We denote this vector by the symbol o. The vector o has the property that

$$z + o = z$$

for all vectors z. The length of the line segment that defines a vector z is called its *modulus*. We denote it by the symbol

$$|z| = l(o, z):$$

see Fig. 4 for a sketch.

We have the following important inequality:

$$|z_1 + z_2| \leqq |z_1| + |z_2|. \tag{5}$$

Formula (5) follows from a consideration of the triangle $o z_1 z_3$ in Fig. 5. The length of the side $o z_1$ is $|z_1|$, the length of the side $z_1 z_3$ is $|z_2|$, and the length of the side $o z_3$ is $|z_1 + z_2|$. Thus the inequality (5) is merely a restatement of the fact that the length of one side of a triangle is less than or equal to the sum of the lengths of the other two sides.

Besides the operations of addition and subtraction, vectors admit a third important operation, multiplication of a vector by a number. Let α be a number and z a vector. We will define the product

$$z' = \alpha z. \tag{6}$$

As in Fig. 6, we draw the line L going through o and z. Suppose that α is a positive number. On the line L, lay off a segment $o z'$ of length $\alpha |z|$ and lying on the same side of 0 as $o z$. The vector z' is the endpoint z' of this line segment. Suppose next that α is negative. In this case, we lay off a segment $o z'$ on the line L with length $|\alpha| \cdot |z|$ and lying on the other side of L from the line segment $o z$. If either α or z is zero, we define the vector αz as o.

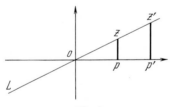

Fig. 6

The vector $z' = \alpha z$ certainly lies on the line L. It is also clear that every vector z' on the line L has the form (6) (the construction has meaning only for $z \neq o$). The required number α is obtained as follows. If the points z' and z lie on the same side of L from the point 0, then the number α is positive and is defined by

$$\alpha = \frac{|z'|}{|z|}.$$

If z' and z lie on opposite sides of o on the line L, then the number α is negative and is defined by

$$\alpha = -\frac{|z'|}{|z|}.$$

We have therefore proved that points of the form (6) $(z \neq o)$ fill out the entire line L when α assumes all possible numerical values.

We now define *the coordinates of a vector*. Since z represents both a point and a vector, it is natural to regard the coordinates of the point z as also the coordinates for the vector z. Thus we write

$$z = (x, y),$$

where z is a vector and x, y are its *coordinates*, that is to say, they are the coordinates of the point z.

Using coordinates we can write our operations on vectors in algebraic form. Recall of course that we have defined these operations purely geometrically. In addition to the vector z, we write two more vectors in coordinate form:

$$z_1 = (x_1, y_1), \qquad z_2 = (x_2, y_2).$$

We then have:

$$z_1 + z_2 = (x_1 + x_2, y_1 + y_2); \tag{7}$$

$$z_1 - z_2 = (x_1 - x_2, y_1 - y_2); \tag{8}$$

$$|z| = +\sqrt{x^2 + y^2}; \tag{9}$$

$$\alpha z = (\alpha x, \alpha y). \tag{10}$$

Let us prove formulas (7)–(10). The reader should turn to Fig. 7. We construct the sum $z_1 + z_2$ in Fig. 7 according to the rule shown in Fig. 5. Draw a horizontal line through z_1 and a vertical line through z_3. Let r be the point where these two lines intersect. The triangles $o p_2 z_2$ and $z_1 r z_3$ are equal to each other. Therefore the length of the line segment $z_1 r$ (with the appropriate sign) is the abscissa of the point z_2, and the length of the line segment $r z_3$ (with the appropriate sign) is the ordinate of the point z_2. This verifies formula (7). The identity (8) follows from (4) and (7). To verify (9), use (3) with $z_2 = z$ and $z_1 = o$. To prove (10), we return to Fig. 6 and consider the two triangles $o p z$ and $o p' z'$. Since they are similar triangles, (10) holds.

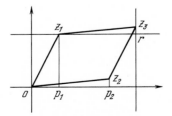

Fig. 7

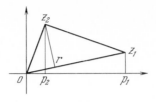

Fig. 8

The distance between two points z_1 and z_2 can be conveniently written in vector notation. Using the identities (3) and (8), we find

$$l(z_1, z_2) = |z_2 - z_1|.$$

Example 1. We will compute the area S of a triangle $o z_1 z_2$, where the points z_1 and z_2 are given by their coordinates as in formula (2). We sketch this triangle in Fig. 8. We will write S in terms of three other areas. 1) First consider the triangle $o p_2 z_2$. Its area S_2 is equal to $\frac{1}{2} x_2 y_2$. 2) Consider next the trapezoid $p_2 p_1 z_1 z_2$. Its area S_3 is equal to $\frac{1}{2}(x_1 - x_2)(y_1 + y_2)$. 3) Finally consider the triangle $o p_1 z_1$, whose area S_1 is equal to $\frac{1}{2} x_1 y_1$. Carrying out a simple calculation, we find

$$S = S_2 - S_1 + S_3 = \tfrac{1}{2}(x_1 y_2 - y_1 x_2), \tag{11}$$

where as already noted we have

$$S_1 = \tfrac{1}{2} x_1 y_1, \ \ S_2 = \tfrac{1}{2} x_2 y_2, \ \text{and} \ \ S_3 = \tfrac{1}{2}(x_1 - x_2)(y_1 + y_2). \tag{12}$$

The calculation just made depends upon fixed positions of the points z_1 and z_2 relative to our coordinate axes. The identity (11) always holds provided that the areas S_1, S_2, and S_3 are given by the formulas (12). The value obtained for the area S is positive provided that the direction $o z_2$ is obtained from the direction $o z_1$ by a counterclockwise rotation. If a clockwise rotation is required to go from $o z_1$ to $o z_2$ within the triangle $o z_1 z_2$, then the formula (11) yields a negative area.

§2. Polar Coordinates

We have introduced a system of Cartesian coordinates in our plane P. That is, we have an axis of abscissas, an axis of ordinates, and the origin of coordinates. With this system there is a closely related system, which is called the system of polar coordinates. Let $z = (x, y)$ be an arbitrary point of the plane P, written in its Cartesian coordinates. We define two different numbers for the point z, which are called its *polar coordinates*. These are the

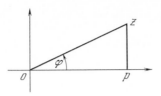

Fig. 9

number ρ, which is equal to the length of the line segment oz, that is, $\rho = l(o, z)$, and the number φ, which is the radian measure of the angle between the positive half-axis of abscissas and the line segment oz. We measure the angle φ in the counterclockwise direction, as indicated in Fig. 9. We will write

$$z = [\rho, \varphi]. \tag{13}$$

The number ρ is called the *radius* of the point z and the number φ its *angle*.

The radius ρ of the point z is uniquely determined. The angle φ on the other hand, is completely undefined if z is the origin. Even if z is not the origin, the angle φ is not uniquely determined. For, if φ is an angle of the point z and k is any integer, the angle

$$\varphi + 2k\pi \tag{14}$$

is also an angle of the point z. In many situations, this point is important. On the other hand, if ρ is any nonnegative number and φ is any number whatever, there is always a point z such that ρ is its radius and φ is its angle. The polar coordinates ρ and φ of a point z do not yield a one-to-one correspondence between coordinates and points of the plane, and to this extent they are less perfect than Cartesian coordinates. Nevertheless, in a variety of cases, they are more convenient than Cartesian coordinates.

It is essential to be able to go back and forth between Cartesian and polar coordinates. For $z = (x, y) = [\rho, \varphi]$, we have the following relations, which may be read off at once from Fig. 9:

$$\left. \begin{array}{l} x = \rho \cos \varphi \\ y = \rho \sin \varphi. \end{array} \right\} \tag{15}$$

These identities are obvious from examining the triangle opz. They have a simple geometric interpretation for points z belonging to the first quadrant. Drop the perpendicular zp from the point z to the axis of abscissas. Since z lies in the first quadrant, its abscissa and ordinate, say x and y, are positive numbers, and we may write

$$x = l(o, p), \qquad y = l(p, z), \qquad \rho = l(o, z).$$

The line segments op and oz are the legs of the right triangle opz, and oz is its hypotenuse. The angle φ of the point z is the angle of this triangle at the vertex o. The identities (15) are the well-known formulas giving the lengths of the legs of a right triangle in terms of the length of its hypotenuse and the acute angle at a vertex. If the point z lies in the 2nd, 3rd, or 4th quadrant, we must pay attention to the signs of the left and right sides in the identities (15). When we do this, we find that (15) holds for all points z in the plane. This is not accidental, but is a consequence of the fact that the rule for choosing the signs of abscissas and ordinates coincides with the rules for choosing signs of cosines and sinces, respectively, in trigonometry. Thus, by the correct use of negative numbers we can carry out calculations without worrying about the the signs of the quantities that we are working with. Had we blundered in our definition of signs, we would have had to use different formulas in (15) for points lying in the four quadrants.

Using negative as well as positive numbers plays an equally important rôle in the distance formula (3). Imagine for a moment that we cannot use negative numbers. Then we would define the abscissa and ordinate of a point z as the lengths of the line segments op and pz, respectively (see Fig. 1), and we would have to specify in some fashion, perhaps by appending one of the numbers 1, 2, 3, 4, which quadrant the point z lies in. In this event, the single formula (3) for the distance between two points would have to be replaced by an array of 16 formulas, to take care of all of the possibilities for z_1 and z_2 to lie in one or another quadrant. Thus each formula would deal with a specific case. The point z_1 lies in one of the quadrants, and this gives 4 possibilities. For each position of z_1, z_2 also can lie in many of the 4 quadrants. Thus there would be a total of 16 possibilities, each of which would require its own distance formula. Similar unpleasant complications would arise with the formulas (15).

Example 2. Consider once again the triangle oz_1z_2 of Example 1, as sketched in Fig. 8. Let us compute its area using polar coordinates

$$z_1 = [\rho_1, \varphi_1], \qquad z_2 = [\rho_2, \varphi_2]. \tag{16}$$

In Fig. 8, let us drop a perpendicular z_2r from the vertex z_2 to the side oz_1 or to its extension. The length of this line segment is the altitude of our triangle. It is equal to $\rho_2 \sin(\varphi_2 - \varphi_1)$. The altitude is to be taken with a fixed sign. The length of the base of our triangle is the number ρ_1, and so we have

$$S = \tfrac{1}{2}\rho_1\rho_2\sin(\varphi_2 - \varphi_1).$$

This is a well-known theorem of trigonometry. Note, however, that we take the area S with a fixed sign, the rule for choosing the sign being as in Example 1. In the equality (11), write the Cartesian coordinates of the points

z_1 and z_2 in terms of their polar coordinates, as in (15). This gives us

$$S = \tfrac{1}{2}\rho_1\rho_2\sin(\varphi_2-\varphi_1) = \tfrac{1}{2}\rho_1\rho_2(\cos\varphi_1\sin\varphi_2 - \sin\varphi_1\cos\varphi_2).$$

We infer that

$$\sin(\varphi_2-\varphi_1) = \sin\varphi_2\cos\varphi_1 - \cos\varphi_2\sin\varphi_1. \tag{17}$$

Thus we have proved one of the fundamental identities of trigonometry, namely, the identity giving the sine of the difference of two angles.

From (17) we can derive three other basic identities of trigonometry: formulas for the sine of a sum, the cosine of a sum, and the cosine of a difference. We need only the following identities, which are easy to verify:

$$\sin(-\varphi) = -\sin\varphi;$$

$$\cos(-\varphi) = \cos\varphi; \tag{18}$$

$$\sin(\tfrac{1}{2}\pi - \varphi) = \cos\varphi.$$

To find the sine of a sum, replace φ_1 by $-\varphi_1$ in (17). We obtain

$$\sin(\varphi_2+\varphi_1) = \sin\varphi_2\cos(-\varphi_1) - \cos\varphi_2\sin(-\varphi_1)$$
$$= \sin\varphi_2\cos\varphi_1 + \cos\varphi_2\sin\varphi_1,$$

or in standard form

$$\sin(\varphi_1+\varphi_2) = \sin\varphi_1\cos\varphi_2 + \cos\varphi_1\sin\varphi_2. \tag{19}$$

Again applying (17) and (18), we find

$$\cos(\varphi_1+\varphi_2) = \sin[(\tfrac{1}{2}\pi-\varphi_2)-\varphi_1]$$
$$= \sin(\tfrac{1}{2}\pi-\varphi_2)\cos\varphi_1 - \cos(\tfrac{1}{2}\pi-\varphi_2)\sin\varphi_1$$
$$= \cos\varphi_2\cos\varphi_1 - \sin\varphi_2\sin\varphi_1.$$

In standard form, this is

$$\cos(\varphi_1+\varphi_2) = \cos\varphi_1\cos\varphi_2 - \sin\varphi_1\sin\varphi_2. \tag{20}$$

Replacing φ_1 by $-\varphi_1$ in (20), we obtain

$$\cos(\varphi_2-\varphi_1) = \cos\varphi_2\cos\varphi_1 + \sin\varphi_2\sin\varphi_1. \tag{21}$$

Example 3. We define the *scalar product* $z_1 \cdot z_2$ of two vectors z_1 and z_2 in terms of their polar coordinates (16). Namely, we write

$$z_1 \cdot z_2 = \rho_1\rho_2\cos(\varphi_2-\varphi_1) = \rho_1\rho_2\cos\gamma, \tag{22}$$

γ being the angle between the vectors z_1 and z_2. The sign of the number γ is of no significance, as (18) shows. The cosine of an acute angle is positive and the cosine of an obtuse angle is negative. From (15) and (21) we obtain the identity

$$z_1 \cdot z_2 = x_1 x_2 + y_1 y_2.$$

Thus we can express the scalar product simply in terms of Cartesian coordinates.

§3. Geometric Representation of Complex Numbers

Cartesian coordinates provide yet another way to connect points and numbers, *viz.*, a geometric representation of complex numbers and operations on complex numbers.

As is known, complex numbers were discovered as a result of the impossibility of finding a real square root of a negative number. Thus one introduces formally a new number i that satisfies the identity

$$i^2 + 1 = 0,$$

or in other terms

$$\sqrt{-1} = \pm i.$$

One then defines a *complex number* z by

$$z = x + iy, \tag{23}$$

where x and y are ordinary numbers, with which we are familiar. We will now designate these numbers as *real numbers*, to distinguish them from complex numbers. We must now define the ordinary operations on complex numbers, namely addition and multiplication and their inverse operations subtraction and division. Everywhere i^2 appears, we replace it by -1. We will find that the arithmetical operations can be carried out on complex numbers just as naturally as on real numbers.

We proceed to formal definitions. Suppose that

$$z_1 = x_1 + iy_1 \quad \text{and} \quad z_2 = x_2 + iy_2$$

are two complex numbers. We define their sum by

$$z_1 + z_2 = (x_1 + x_2) + i(y_1 + y_2). \tag{24}$$

We define their product by

$$z_1 z_2 = (x_1 + i y_1)(x_2 + i y_2) = x_1 x_2 + i x_1 y_2 + i y_1 x_2 + i^2 y_1 y_2$$
$$= (x_1 x_2 - y_1 y_2) + i(x_1 y_2 + y_1 x_2). \qquad (25)$$

Nothing could be more natural than to represent a complex number z represented as in (23) by a point in the plane P, namely, the point with coordinates (x, y) or the vector with coordinates (x, y): $z = (x, y)$. We will call the plane P in which complex numbers z are represented by points *the complex plane* or *the plane of the complex variable z.*

It is reasonable to call complex numbers z of the form

$$z = x + i0 = x$$

real numbers. All of them lie on the axis of abscissas, and so we call the axis of abscissas in the complex plane *the real axis.* Complex numbers of the form

$$z = 0 + iy = iy$$

are called *pure imaginaries* or simply *imaginary.* All of them lie on the axis of ordinates, which we will call *the imaginary axis.* The complex number $z = 0 + i0 = 0$ coincides with the origin. The real numbers $+1$ and -1 lie on the real axis to the right and left of zero, respectively, at distance one from zero. The imaginary numbers $+i$ and $-i$ lie on the imaginary axis above and below zero, respectively, both at distance one from zero.

Let us now write the complex number z in terms of the polar coordinates of the point z. Formula (15) gives us

$$x = \rho \cos \varphi \quad \text{and} \quad y = \rho \sin \varphi.$$

Thus the complex number $z = x + iy$ can be written as

$$z = \rho(\cos \varphi + i \sin \varphi). \qquad (26)$$

We call this identity *the trigonometric representation of the complex number z.* The nonnegative number ρ is called the *modulus* of the complex number z and the angle φ its *argument.* The number ρ is uniquely determined by z and is denoted by the symbol

$$|z| = \rho = +\sqrt{x^2 + y^2}.$$

We should emphasize once again that the argument φ of z is not defined uniquely by z. In fact, for $|z| = 0$ the argument is not defined at all. For $|x| \neq 0$ and for each argument φ of z, all of the numbers

$$\varphi + 2k\pi \qquad (27)$$

are also arguments of z (here k can be any integer). Compare this observation with formula (14).

We now give geometric interpretations of the definitions (24) and (25) for addition and multiplication of complex numbers.

The two definitions (7) and (24) show that addition of two complex numbers z_1 and z_2 corresponds to addition of their vectors z_1 and z_2. That is, the sum $z_3 = z_1 + z_2$ of the vectors represents the sum z_1 and z_2 of the complex numbers. As noted in (4), the difference of two vectors is always defined. Hence we can also subtract one complex number from another. This geometric representation of addition gives us the important inequality

$$|z_1 + z_2| \leqq |z_1| + |z_2|, \tag{28}$$

which is equivalent to the inequality (5).

To give a geometric interpretation of the product of two complex numbers z_1 and z_2, let us write these numbers as well as their product w in trigonometric form:

$$\left.\begin{array}{l} z_1 = \rho_1(\cos \varphi_1 + i \sin \varphi_1); \\[4pt] z_2 = \rho_2(\cos \varphi_2 + i \sin \varphi_2); \\[4pt] z_1 z_2 = w = \sigma(\cos \psi + i \sin \psi). \end{array}\right\} \tag{29}$$

In computing the product in trigonometric form, we use the addition formulas (19) and (20). Combining these with (25) and (19), we find

$$\begin{aligned} w = z_1 z_2 &= \rho_1 \rho_2 [(\cos \varphi_1 \cos \varphi_2 - \sin \varphi_1 \sin \varphi_2) \\ &\quad + i(\sin \varphi_1 \cos \varphi_2 + \cos \varphi_1 \sin \varphi_2)] \\ &= \rho_1 \rho_2 [\cos(\varphi_1 + \varphi_2) + i \sin(\varphi_1 + \varphi_2)]. \end{aligned}$$

Combining this with the last line of (29), we write

$$\sigma(\cos \psi + i \sin \psi) = \rho_1 \rho_2 [\cos(\varphi_1 + \varphi_2) + i \sin(\varphi_1 + \varphi_2)]. \tag{30}$$

We have thus shown that

$$\sigma = \rho_1 \rho_2 \quad \text{and} \quad \psi = \varphi_1 + \varphi_2. \tag{31}$$

This means that when we multiply two complex numbers, we multiply their moduli ρ_1 and ρ_2 and add their arguments φ_1 and φ_2.

From (31) it is obvious that the quotient of two complex numbers exists provided only that the denominator is not zero. Let w and z_1 be complex numbers. We wish to define z_2 so that $w = z_1 z_2$. Formula (31) shows that

$$\rho_2 = \frac{\sigma}{\rho_1} \quad \text{and} \quad \varphi_2 = \psi - \varphi_1.$$

Formula (30) for the product of two complex numbers can be extended to any number of factors: simply use (30) over and over. Suppose that

$$z_j = \rho_j(\cos \varphi_j + i \sin \varphi_j) \quad (j = 1, 2, \ldots, n)$$

is a sequence of complex numbers, and write $w = z_1 z_2 \cdots z_n$. We find that

$$w = \sigma(\cos \psi + i \sin \psi)$$
$$= \rho_1 \cdots \rho_n [\cos(\varphi_1 + \cdots + \varphi_n) + i \sin(\varphi_1 + \cdots + \varphi_n)].$$

In particular, if all of the numbers $z_1, \cdots, z_n$ are equal to a fixed number z, we find the following interesting result:

$$z^n = \sigma(\cos \psi + i \sin \psi) = \rho^n(\cos n\varphi + i \sin n\varphi). \tag{32}$$

With every complex number $z = x + iy$, we associate its *complex conjugate* number $x - iy$, written briefly as $\bar{z}$:

$$\bar{z} = x - iy.$$

Complex conjugates are often very useful. If we write our complex number in trigonometric form as in (26), the complex conjugate can be written as

$$\bar{z} = \rho[\cos(-\varphi) + i \sin(-\varphi)]. \tag{33}$$

Thus the modulus of the complex conjugate is the same as the modulus of the number itself, while the argument changes its sign.

It is obvious that $\bar{\bar{z}} = z$, that is, applying conjugation twice brings us back to the complex number we started with. Note too that a complex number is equal to its own conjugate if and only if it is real.

Geometrically, we form the complex conjugate $\bar{z}$ by reflecting z in the real axis.

From the identities (33), (24), and (30) it is obvious that

$$\overline{z_1 + z_2} = \bar{z}_1 + \bar{z}_2 \quad \text{and} \quad \overline{z_1 z_2} = \bar{z}_1 \bar{z}_2.$$

These identities obviously hold for any number of summands or factors. For example, we have $\overline{z^k} = (\bar{z})^k$. If a is a real number, then

$$\overline{a z^n} = \bar{a}(\bar{z})^n = a(\bar{z})^n.$$

For an arbitrary polynomial

$$f(z) = a_0 z^n + a_1 z^{n-1} + \cdots + a_n$$

with real coefficients, we find that

$$\overline{f(z)} = \overline{a_0 z^n + a_1 z^{n-1} + \cdots + a_n} = \overline{a_0 z^n} + \overline{a_1 z^{n-1}} + \cdots + \overline{a}_n$$
$$= a_0(\overline{z})^n + a_{n-1}(\overline{z})^{n-1} + \cdots + a_n = f(\overline{z}).$$

That is, for a polynomial with real coefficients we have

$$\overline{f(z)} = f(\overline{z}). \tag{34}$$

The identity (34) has a most important consequence. Suppose that γ is a root of a polynomial $f(z)$ with real coefficients, that is,

$$f(\gamma) = 0.$$

Then the complex conjugate $\overline{\gamma}$ is also a root of $f(z)$: $f(\overline{\gamma}) = 0$. This follows from (34):

$$f(\overline{\gamma}) = \overline{f(\gamma)} - \overline{0} = 0.$$

Example 4. Let us now use the identity (32) to find the nth roots of an arbitrary nonzero complex number w. (The number 0 is uninteresting for this question, since all of its nth roots are 0.) We must solve the equation

$$z^n = w, \tag{35}$$

where z is the unknown and w is a given nonzero complex number.

Consider first the case $w = 1$. Using (32), we rewrite (35) in the form

$$\rho^n(\cos n\varphi + i \sin n\varphi) = \cos 2k\pi + i \sin 2k\pi, \tag{36}$$

where k is an arbitrary integer (see (27)). Solving (36), we see that

$$\rho = 1 \quad \text{and} \quad \varphi = \frac{2k\pi}{n},$$

k still being an arbitrary integer. In other words, every n^{th} root of unity has the form

$$\sqrt[n]{1} = \cos\frac{2k\pi}{n} + i \sin\frac{2k\pi}{n},$$

where k is an integer. It would appear that the last formula gives us an infinite number of nth roots of unity, but in fact there are only n of them. To see this, select a particular n^{th} root of unity:

$$\varepsilon = \cos\frac{2\pi}{n} + i \sin\frac{2\pi}{n}.$$

Plainly an arbitrary n^{th} root of unity has the form

$$\cos\frac{2k\pi}{n}+i\sin\frac{2k\pi}{n}=\varepsilon^{k}. \tag{37}$$

Now take two distinct values for k, say p and q. The left side of (37) gives the same number if and only if

$$\frac{2p\pi}{n}-\frac{2q\pi}{n}=2l\pi,$$

where l is an integer. Therefore we get all distinct numbers (37) if we give to k in turn the values $0, 1, 2, \ldots, n-1$. That is, the numbers

$$\varepsilon^{0}, \varepsilon^{1}, \ldots, \varepsilon^{n-1} \tag{38}$$

are all of the distinct n^{th} roots of unity. From (37), we see that the numbers (38) all lie on a circle of radius one and center at the origin. They are the vertices of a regular polygon inscribed in this circle that has n sides. Furthermore, one of these vertices is the point $\varepsilon^{0}=1$, lying on the real axis. In Figs. 10, 11, and 12 we sketch the 3rd, 4th, and 6th roots of unity.

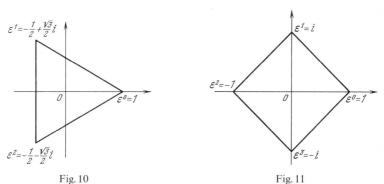

Fig. 10   Fig. 11

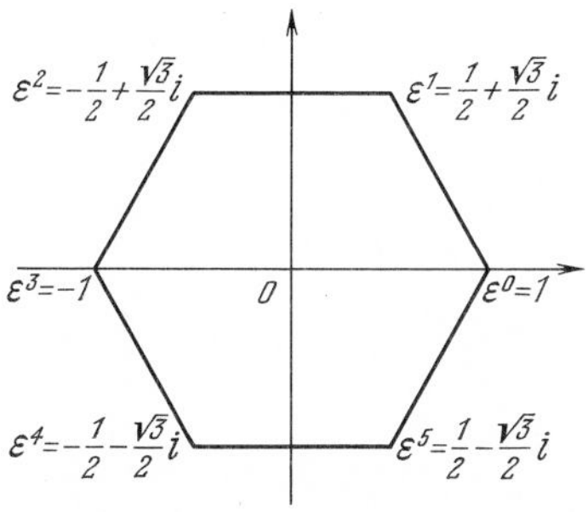

Fig. 12

We will now describe the nth roots of an arbitrary nonzero complex number w. Recall that the argument of a complex number is defined only up to a summand of the form $2k\pi$, where k is an integer. Bearing this in mind, we choose some specific argument ψ_0 for the number w. We write the equation (35) in trigonometric form:

$$\rho^n(\cos n\varphi + i\sin n\varphi) = [\cos(\psi_0 + 2k\pi) + i\sin(\psi_0 + 2k\pi)],$$

where k is an arbitrary integer. From this we obtain

$$\rho = \sqrt[n]{\sigma}$$

where ρ is a positive number, and we also obtain

$$\varphi = \frac{\psi_0}{n} + \frac{2k\pi}{n},$$

k still being an arbitrary integer. Let us write

$$z_0 = \rho\left(\cos\frac{\psi_0}{n} + i\sin\frac{\psi_0}{n}\right).$$

We find that an arbitrary n^{th} root of w has the form

$$\sqrt[n]{w} = z_0\,\varepsilon^k,$$

where k is one of the numbers $0, 1, \ldots, n-1$, and z_0 is a fixed n^{th} root of w. That is, the nth roots of w are the vertices of a regular n-sided polygon inscribed in the circle of radius ρ and center at the origin. One of the vertices of this polygon is the number z_0.

Example 5. The relation

$$i^2 = -1 \tag{39}$$

is a pure definition. It describes the new number i, and so it cannot be proved. It is simply an agreement.

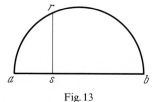

Fig. 13

Notwithstanding, there have been people who wished to prove (39). They have gone about the task as follows. In Fig. 13, we have sketched a semicircle with its diameter ab. From an arbitrary point r of the semicircle, drop a perpendicular rs to the diameter ab. By a classical theorem of elementary geometry, the length of rs is the geometric mean of the lengths of the intervals as and sb. Although we are speaking of lengths, we will commit no serious blunder in saying that the square of the interval rs is equal to the product of the intervals as and bs.

We now consider the complex plane. Let a be the point -1, let b be the point $+1$, and let r be the point i. Plainly s is the point 0. The author of the "proof" of (39) reasons as follows. The interval rs is i, the interval as is -1, and the interval sb is $+1$. The theorem just cited from elementary geometry yields the equalities $i^2 = (-1)(+1) = -1$.

Despite the complete senselessness of this "proof", it was published in Kol'man's book „*The Object and Method of Mathematics*." Even more, it was set forth in mathematical lessons in the schools of Moscow: and this up to the second world war.

Supplement to Chapter I

In this supplement, we construct Cartesian coordinates in space and define vectors in space and operations on them. That is, we extend the previous work in the plane to space. We give the proof in less detail than for the case of the plane in the expectation that the reader will supply the missing details.

1. Coordinates in Space. Suppose that we are given a certain line R and a certain point x in space. Through the point x, we construct a plane Q that is perpendicular to the line R. The point p where the line R intersects the plane Q is called the projection of x onto R.

In §1, we described the rectangular (Cartesian) system of coordinates in a plane P. We will extend this to a system of rectangular coordinates in space. Suppose that we have a plane P in space, and suppose also that it is horizontal. Introduce a system of Cartesian coordinates in the plane P. We will call the axis of abscissas the first axis and the axis of ordinates the second axis. We take as our third axis a line through the origin o in P and perpendicular to the plane P. We orient this axis by going from below P to above P. Thus a rectangular system of Cartesian coordinates in space is defined by three axes. Each of them is a line with a specified direction on it. All pairs of lines are mutually perpendicular, and all three go through a common point o, called *the origin of coordinates* or *the origin*. Through each pair of axes we pass a coordinate plane. The plane through the first and second axes is designed as the coordinate plane $(1,2)$; the coordinate planes $(2,3)$ and $(1,3)$ are defined similarly.

Having chosen our coordinates axes, we associate with each point x of space a triple of numbers x_1, x_2, x_3, called its *coordinates*. Let p_1, p_2, and p_3 be the projections of x onto the first, second, and third axes, respectively. Let x_1 be the length of the line segment op_1 if p_1 is in the positive direction from o on the first axis. Let x_1 be the negative of the length of op_1 if p_1 lies in the negative direction from o on the first axis. The second and third coordinates x_2 and x_3 are defined in just the same way from op_2 and op_3, respectively. We write

$$x = (x_1, x_2, x_3). \tag{40}$$

It is also clear that if we choose the numbers x_1, x_2, x_3 arbitrarily, then there is exactly one point x in space with coordinates x_1, x_2, x_3. That is, the relation (40) yields a one-to-one correspondence between points of space and all triples of numbers (x_1, x_2, x_3).

Now let $y = (y_1, y_2, y_3)$ be any other point in space. The *distance* $l(x, y)$ between the points x and y is easily proved to be equal to

$$l(x, y) = +\sqrt{(y_1 - x_1)^2 + (y_2 - x_2)^2 + (y_3 - x_3)^2}. \tag{41}$$

2. Vectors in Space. We proceed by analogy with the plane. Let a and b be any two points in space. The line segment starting at a and ending at b is called a *vector*. The point a is called the *initial point of the vector ab* and the point b is called its *terminal point*. Two vectors ab and cd are said to be *equal* if they lie in a single plane and are equal considered as vectors in this plane (see §1 of the main text). For every vector ab in space, there is exactly one vector ox with initial point o that is equal to ab. We designate this vector by the single letter x. The coordinates of the point x are regarded as the coordinates of the vector ox, or x. We express this by the equality

$$x = (x_1, x_2, x_3).$$

The length of the vector x is called its *modulus* and is denoted by the symbol $|x|$. Applying the equality (41), we see that

$$|x| = +\sqrt{x_1^2 + x_2^2 + x_3^2}.$$

Let x and y be any two vectors in space (both have initial point o). Let Q be any plane in space that contains o, x, and y. We define the sum $x + y$ of the vectors x and y by the rules given in §1 for the sum of two vectors in the plane Q. In the same way we define the product of the vector x by any real number α; that is, we obtain the vector αx. We can also define $x + y$ and αx without passing a plane through o, x, and y. In fact, one proves without difficulty that if

$$x = (x_1, x_2, x_3) \quad \text{and} \quad y = (y_1, y_2, y_3),$$

then

$$x+y=(x_1+y_1, x_2+y_2, x_3+y_3)$$

and

$$\alpha x=(\alpha x_1, \alpha x_2, \alpha x_3).$$

For x and y any two vectors in space, we define their *scalar product* $x \cdot y$ by

$$x \cdot y = x_1 y_1 + x_2 y_2 + x_3 y_3.$$

It is clear that

$$x \cdot x = x^2 = |x|^2,$$

that

$$x \cdot y = y \cdot x,$$

and that

$$(\alpha x) \cdot y = \alpha(x \cdot y).$$

Furthermore, if x, y, and z are any three vectors in space, then we have

$$(x+y) \cdot z = x \cdot z + y \cdot z. \tag{42}$$

We now prove the fundamental equality

$$x \cdot y = |x| \cdot |y| \cos \gamma, \tag{43}$$

where γ is the angle between the vectors x and y. We have already proved the equality (43) for vectors in the coordinate plane $(1, 2)$, that is, the plane P. For arbitrary vectors x and y, we use (42) to write

$$(x+y)^2 = x^2 + 2x \cdot y + y^2,$$

or

$$x \cdot y = \tfrac{1}{2}[(x+y)^2 - x^2 - y^2]. \tag{44}$$

the identity (44) holds in particular for vectors x and y that lie in the plane P. For such vectors, the left side of (44) is equal to $|x| \cdot |y| \cos \gamma$ (see formula (22) of the main text).

Thus for vectors x and y lying in the plane P, we have

$$|x| \cdot |y| \cos \gamma = \tfrac{1}{2}[(x+y)^2 - x^2 - y^2]. \tag{45}$$

The identity (45) does not refer to individual coordinates but only to the lengths of the vectors x, y, and $x+y$, and to the cosine of the angle between them. We proved (45) for vectors x and y lying in the coordinate plane P.

The same proof is valid for vectors x and y that lie in an arbitrary plane Q through the origin o. Therefore (45) holds for any two vectors x and y, since there is at least one plane that contains o, x, and y. We have verified (44) and (45) for arbitrary vectors x and y. Since their right sides coincide, their left sides must be equal. This proves (43) for all pairs of vectors x and y.

The identity (43) gives a simple test for perpendicularity of two vectors. In fact, x and y are perpendicular if and only if

$$x \cdot y = 0.$$

Let e be an arbitrary vector of length 1. Draw the line R that contains both o and e. Every point x' (vector x') lying on R has the form

$$x' = \alpha\, e. \tag{46}$$

We may regard α as the coordinate of the point x' on the line R: we take R with the direction from 0 to e. Now let x be any vector (point) in space, and let x' be the projection of x onto the line R. From the identity (43), we see that

$$x' = (e \cdot x)\, e. \tag{47}$$

Chapter II
Coordinates and Lines in the Plane

In Chapter I, we studied primarily individual points in the plane, or at most finite sets of points: for example, the vertices of triangles, parallelograms, and regular polygons. The sole exception was a straight line, which we encountered in defining the product of a vector by a number. In this second chapter, we take up the description of lines in coordinate form. We will first show that the simplest possible relation between coordinates x and y yields a line. We then define the graph of a function and functions themselves. We will also obtain the equation of a line from the geometric properties of lines. We will do the same for the circle, the ellipse, the hyperbola, and the parabola, defining these curves first by their geometric properties. After this, we take up the question of defining a line parametrically. We define the winding number of a closed curve with respect to the origin. In the last section of this chapter, we present a geometric interpretation of the behavior of polynomials with complex coefficients. Using the winding number, we prove the fundamental theorem of algebra: a polynomial of degree n admits n roots.

§4. Graphs of Functions and Functions

Let us write the following equality:

$$f(x) = a_0 x^n + a_1 x^{n-1} + \cdots + a_n. \tag{1}$$

On the right side of (1) we have written a polynomial of degree n in the variable x and with coefficients $a_0, a_1, \ldots, a_n$. We suppose that these coefficients are real numbers. On the left we see the symbol $f(x)$, which we will first and foremost regard as an abbreviation for the polynomial on the right side of (1). That is, instead of saying "the polynomial $a_0 x^n + a_1 x^{n-1} + \cdots + a_n$", we may simply say "the polynomial $f(x)$." For example, we may refer to the equation $f(x) = 0$, with the understanding that $f(x)$ is the polynomial on the right side of (1). We may also write one number or another instead of x in (1) and again obtain equalities. For example, setting $x = 0$, we obtain

$f(0)=a_n$, and setting $x=1$, we obtain $f(1)=a_0+a_1+\cdots+a_n$. Thus, if we know the coefficients $a_0, a_1, \ldots, a_n$, we can compute the value of the polynomial $f(x)$ for an arbitrary value of x simply by substituting this numerical value in the right side of (1). Since x can assume all numerical values, we say that x is *an independent variable*. Since we can compute the numerical value of $f(x)$ as soon as we know x, we say that $f(x)$ *is a function of the independent variable* x. The variable x is also sometimes called *the argument of the function* $f(x)$. In the case we are considering, $f(x)$ is the polynomial described in (1). There are many other rules for describing functions $f(x)$. Some of these are expressed in one or another formula, or by other means, perhaps in words. Thus we arrive at the concept of a function. *A function (of the independent variable* x*)* is a quantity $f(x)$ that one can identify as soon as one knows the value of x.

For the sake of definiteness, let us suppose for the nonce that $f(x)$ is defined by (1).

We consider the equation

$$y=f(x). \tag{2}$$

When we assign various values to the independent variable x, the equality (2) will give us completely determined numerical values for y. For this reason, y is called *a dependent variable*. Let us choose a system of Cartesian coordinates in a certain plane P, as was done in §1. In this plane, we consider all points

$$z=(x,y), \tag{3}$$

where x can be arbitrary and y is computed for each x by the formula (2). The result of doing this is to obtain a curve in the plane P. This curve is the set of all points $z=(x,y)$, as x runs through all possible values and the corresponding points y are computed according to (2). This curve is called *the graph of the function* $f(x)$. We also say that the curve is defined by the equation (2). Thus the equation (2) defines a curve in the plane P. We denote this curve by the symbol K. The point (3) on the curve can be written in the form

$$z=(x,f(x)), \tag{4}$$

an expression that emphasizes the fact that the ordinate of the point is a function of its abscissa.

Sometimes the independent variable is denoted by a letter different from x, for example t. This is common in case the independent variable is time. Thus we can consider a function $\varphi(t)$ of the independent variable t and consider the equation

$$\varphi=\varphi(t). \tag{5}$$

Here we use the same letter φ to denote the independent variable and also the expression $\varphi(t)$ which defines the function. This is done not only to economize on the number of letters we use but also to emphasize that the dependent variable φ is defined by the function $\varphi(t)$. The fact that we write the independent and dependent variables not as x and y but as t and φ does not deprive us of the possibility of describing the function $\varphi(t)$ geometrically, by means of a graph. To construct the graph of the function $\varphi(t)$ defined by (5), we must take the quantity t on the axis of abscissas and the quantity $\varphi(t)$ as in (5) on the axis of ordinates. We say that we lay off t on the axis of abscissas and φ on the axis of ordinates.

If we contemplate for a while the definition of the graph of a function, we involuntarily encounter doubts. In fact, x assumes an infinite set of values, and it is totally impossible to compute all of the corresponding values of y. Of course, one can choose a certain finite number of values of x and compute the corresponding values of y. Thus we can find a certain finite number of points on the curve K and thus get some idea of its shape. In spite of this deficiency, we say that the equation (2) defines the curve consisting of all points of the form (4).

In elementary geometry a curve is usually considered to be the path of a moving point. For example, a line is described by the end of a pencil that moves along a straightedge. A circle is described by one tip of a pair of moving compasses, the other tip being fixed. The same intuitive notion can be applied to give some idea of the curve K defined by equation (2). Turn back to Fig. 2 and imagine that the abscissa x starts at a and moves along the axis of abscissas to the point b, where a is less than b. Thus the point labelled p in Fig. 2 moves from left to right on the axis of abscissas, and the value of the ordinate $y = f(x)$ lies on the perpendicular $p\,z$. Thus, during the time that p is moving, the point x traces out a piece of the curve K.

We now take up some special cases of the polynomial $f(x)$ in (1) and construct the corresponding curves K.

Suppose that

$$f(x) = \alpha x^2,$$

where α is a positive number. The equation (2) then has the form

$$y = \alpha x^2. \tag{6}$$

This equation defines the curve K in our plane P that consists of all points of the form

$$z = (x, \alpha x^2),$$

where x is an arbitrary real number. We can say immediately that our curve K lies above the axis of abscissas, since its ordinates αx^2 are always nonnegative. It also goes through the origin, since $y = 0$ for $x = 0$. Furthermore, the curve K is symmetric with respect to the axis of ordinates, since $(-x, y)$ lies on K if (x, y) does. Thus, when we know how K looks in the first quadrant, we can reflect this part of K in the axis of ordinates and

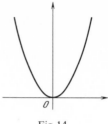

Fig. 14

obtain the form of K in the second quadrant. From the equation (6) it is clear that as x increases from 0, y also increases. At first the increase in y is slow, but then it grows faster and faster. That is, the curve K looks more or less as sketched in Fig. 14.

Let us now consider the curve K defined by the equation

$$y = x^3 + \beta x, \tag{7}$$

where β is a specified number. We must consider two cases here: 1) β is nonnegative; 2) β is negative. In case 1), the function y increases with increasing x. It is also clear that y is negative if x is negative and positive if x is positive. If (x, y) is a point on the curve, then $(-x, -y)$ is also a point on the curve. That is, the curve K is symmetric with respect to the origin. The curve goes through the first and the third quadrants. To sketch the curve, we need only determine its behavior in the first quadrant, since we can then use symmetry to sketch it also in the third quadrant. As x increases, beginning with zero, y also increases, at first comparatively slowly, but then faster and faster. Thus the curve has the general form sketched in Fig. 15.

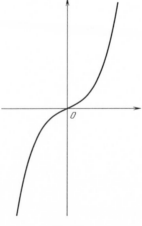

Fig. 15

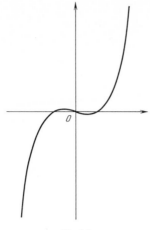

Fig. 16

For negative values of the constant β, the behavior of the curve K is more complicated. Plainly the curve is still symmetric with respect to the origin, but it now has not one point of intersection with axis of abscissas, but three. To find these points, we set y equal to zero in the equation (7). Thus we obtain the values of x where the curve crosses the axis of abscissas. The equation

$$x^3 + \beta x = 0$$

can at once be solved for x: we get $x = 0$ and $x = \pm\sqrt{-\beta}$. In Fig. 16, we sketch the curve K for negative values of β.

The curve K defined by the equation (6) is called *a parabola*. The curve K defined by the equation (7) is called *a cubic parabola*.

We now take up an example of a function $f(x)$ that is not a polynomial. Suppose that

$$f(x) = \sin x.$$

Then our curve K is defined by the equation

$$y = \sin x. \tag{8}$$

For every integer k and all numbers x, we have $\sin(x + 2k\pi) = \sin x$. Therefore to construct the curve K it suffices to study it only in the interval where x runs from 0 to 2π. The entire curve K is obtained from this piece K' of the curve by sliding K' along the axis of abscissas by amounts $2k\pi$, where k is an arbitrary integer. It is easy to sketch the piece K' from equation (8). First the point (x, y) moves in the first quadrant. As x increases from 0 to $\frac{1}{2}\pi$, the quantity y increases from 0 to 1. As x increases from $\frac{1}{2}\pi$

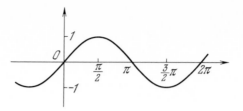

Fig. 17

to π, the quantity y decreases from 1 to 0. Next, as x increases from π to $\frac{3}{2}\pi$, the quantity y decreases from 0 to -1. As x increases from $\frac{3}{2}\pi$ to 2π, the quantity y increases from -1 to 0. The entire curve K has the form sketched in Fig. 17. It is called *the sine curve* or simply *the sine*.

Example 1. We now consider the curve in the plane defined by the equation

$$y=\frac{\gamma}{x}, \tag{9}$$

where γ is a positive number. Note first of all that the equation (9) assigns no value y to the number $x=0$, since division by zero is impossible. We therefore agree that the function

$$f(x)=\frac{\gamma}{x} \tag{10}$$

is undefined at $x=0$. In mathematics we are compelled to study functions of this sort, which are not defined for all values of the argument x. We can also take the point of view that the equation (9) defines two distinct curves in the plane P: one curve for $x>0$ and another for $x<0$. The first of these lies in the first quadrant and the second in the third quadrant. If a point (x, y) lies on the first curve, then the point $(-x, -y)$ lies on the second. That is, the second curve is obtained from the first by a symmetric mapping through the origin. Thus we easily construct the second curve once we know the first one. The point $\left(x,\dfrac{\gamma}{x}\right)$ on the first curve gets arbitrarily close to the axis of abscissas when x grows very large. Also this point gets arbitrarily close to the axis of ordinates when x goes to zero, and the value of y grows without any bound. We sketch both curves in Fig. 18. Plainly we have two distinct curves. Despite this, we frequently regard the equation (9) as defining a single curve. We say that the function (10) has a *discontinuity* or *jump* at the point $x=0$.

We now describe another function that has a discontinuity at $x=0$. This function plays an important rôle in mathematics, and so we give it a special

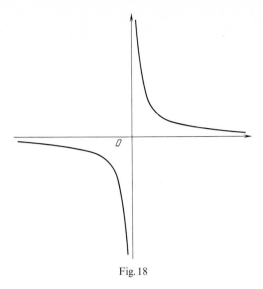

Fig. 18

name:

$$f(x) = \text{sign } x.$$

We define this function as follows:

$$\text{sign } x = +1 \quad \text{for } x > 0;$$

$$\text{sign } x = -1 \quad \text{for } x < 0.$$

We do not define the function sign x for $x = 0$. Thus the function sign x is defined by two different formulas, one for $x > 0$ and a different one for $x < 0$. Such cases of defining functions by distinct formulas in various intervals are not uncommon in mathematics. The equation

$$y = \text{sign } x \qquad\qquad (11)$$

produces a curve in our plane P consisting of two pieces of distinct forms. For $x > 0$ we obtain a straight line parallel to the axis of abscissas and at unit distance above this axis. For $x < 0$, we obtain a straight line parallel to the axis of abscissas and at unit distance below this axis. See Fig. 19. From this point of view, the equation (11) defines two different curves in the plane P. Nevertheless, we regard the equation (11) as defining a single curve. In the same sense, we regard the function sign x as a **single** function, which undergoes a discontinuity or jump at the point 0.

The two functions $\dfrac{y}{x}$ and sign x, both undefined at the point $x = 0$, have "breaks" at this point. The graphs of these functions cannot be regarded as

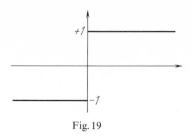

Fig. 19

the path traced out by a moving point in the plane. Thus these functions display certain kinds of breaks, or what we may call violations of continuity. Our initial concept is that of continuity of a function. Later on we will give a precise definition of the notion of continuity of a function. For the time being we will use it intuitively, connecting it with the idea of motion.

§5. Ellipses, Hyperbolas, and Parabolas

The main point of §4 was to construct a curve from a function defined in some fashion. That is, we began algebraically and obtained a geometric object. In this section, we follow a different path. We will consider geometric properties of certain curves and derive from these properties equation for the curves. We will do this for circles with arbitrary centers and for ellipses, hyperbolas, and parabolas which we place in particular positions with respect to a system of coordinates.

Consider first a circle K in the plane P having center at the point z_0 $=(x_0, y_0)$ and radius r. Recall that x_0 and y_0 are the coordinates of the point z_0. A point z in the plane P belongs to *the circle K* if and only if its distance from z_0 is equal to r. We write this as

$$l(z_0, z) = r.$$

Using formula (3) of Chapter I, we rewrite this formula as

$$+\sqrt{(x-x_0)^2 + (y-y_0)^2} = r.$$

We eliminate the radical by squaring both sides of the last equality. This gives us

$$(x-x_0)^2 + (y-y_0)^2 = r^2. \tag{12}$$

This is *an equation of the circle K.* If the point z_0, the center of the circle, happens to be the origin, then (12) assumes a particularly simple form:

$$x^2 + y^2 = r^2. \tag{13}$$

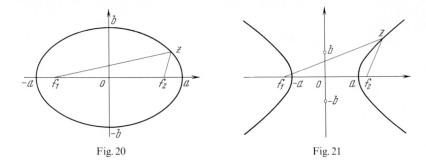

Fig. 20   Fig. 21

We will now find equations for ellipses and hyperbolas. The needed calculations can be carried out simultaneously. Let us first give the geometric definitions of both curves. Both curves admit *foci*, which are simply two points f_1 and f_2 in the plane P.

An ellipse is defined by the following condition. A point z lies on an ellipse if and only if

$$l(z, f_1) + l(z, f_2) = 2a, \tag{14}$$

where a is a fixed positive number. The geometrical situation is sketched in Fig. 20. In words, we say that a point z lies on our ellipse if and only if the sum of its distances from the two foci is constant. We admit the possibility that f_1 and f_2 are the same point, in which case the ellipse is simply the circle with radius a and center f_1.

An hyperbola is defined by the following condition. A point z lies on an hyperbola if and only if

$$l(z, f_1) - l(z, f_2) = 2a, \tag{15}$$

where again a is a fixed positive number. We sketch the situation in Fig. 21. In words, we say that a point z lies on an hyperbola if and only if the difference of its distances from the two foci is a constant. If we interchange the rôles of f_1 and f_2, we get a condition different from (15):

$$l(z, f_2) - l(z, f_1) = 2a. \tag{16}$$

The relations (15) and (16) actually define two different curves. However, when we work out these relations in terms of coordinates, they become the same so far as the equations are concerned. It is therefore appropriate to write a description of the hyperbola as follows:

$$|l(z, f_1) - l(z, f_2)| = 2a. \tag{17}$$

We wish to make our computations for both the ellipse and the hyperbola simultaneously, and for this reason we combine (14) and (17) into the single condition

$$|l(z, f_1) + \varepsilon\, l(z, f_2)| = 2a, \tag{18}$$

where ε can be either $+1$ or -1.

It is convenient to choose the coordinate axes as follows. The axis of abscissas goes through the foci f_1 and f_2 in the direction from f_1 to f_2, and the origin o is taken midway between f_1 and f_2. The axis of ordinates then has to be a perpendicular to the axis of abscissas through the point o. We draw our picture so that the axis of abscissas goes horizontally from left to right, while the axis of ordinates goes from below the axis of abscissas to above it. Let f be the common length of the intervals of_1 and of_2. With this system of coordinates, we have

$$f_1 = (-f, 0), \quad f_2 = (f, 0).$$

Formula (3) of Chapter I shows that for $z = (x, y)$, we have

$$l(z, f_1) = +\sqrt{(x+f)^2 + y^2} \tag{19}$$

and

$$l(z, f_2) = +\sqrt{(x-f)^2 + y^2}. \tag{20}$$

To shorten our computations, we define

$$u = x^2 + y^2 + f^2 \quad \text{and} \quad v = xf. \tag{21}$$

With this notation, formulas (19) and (20) assume the forms

$$l(z, f_1) = +\sqrt{u + 2v}, \tag{22}$$

$$l(z, f_2) = +\sqrt{u - 2v}. \tag{23}$$

We wish to eliminate the radicals in equation (18). To do this, we first square both sides of (18). Applying (22) and (23), we write the result of this as

$$u + 2v + 2\varepsilon\sqrt{u^2 - 4v^2} + u - 2v = 4a^2.$$

A bit of algebra reduces this equation to the form

$$+\varepsilon\sqrt{u^2 - 4v^2} = 2a^2 - u.$$

Squaring both sides of the last equation, we obtain

$$u^2 - 4v^2 = 4a^4 - 4a^2 u + u^2.$$

A simple manipulation of this equality gives us

$$v^2 = a^2 u - a^4.$$

In the last formula, replace u and v by their values from (21). We obtain

$$x^2(a^2 - f^2) + y^2 a^2 = a^4 - a^2 f^2.$$

Under the hypothesis that f be different from a, we can divide the last equality by $a^2(a^2 - f^2)$ to obtain the equation

$$\frac{x^2}{a^2} + \frac{y^2}{a^2 - f^2} = 1. \tag{24}$$

We must now treat separately the cases $f < a$ and $f > a$. In the first case, we write $b^2 = a^2 - f^2$ and in the second we write $b^2 = f^2 - a^2$. In both cases we take b to be a positive number. In the first case we obtain the equation

$$\frac{x^2}{a^2} + \frac{y^2}{b^2} = 1, \tag{25}$$

and in the second case we obtain the equation

$$\frac{x^2}{a^2} - \frac{y^2}{b^2} = 1. \tag{26}$$

Equation (25) is *an equation for an ellipse*, while equation (26) is *an equation for an hyperbola*.

At first glance, it may seem curious that the number $\varepsilon = \pm 1$, which produces an ellipse for $\varepsilon = +1$ and an hyperbola for $\varepsilon = -1$, has vanished as a result of our calculations. The difference between the ellipse and the hyperbola is expressed differently, as we have seen. The case $f < a$ gives the ellipse and the case $f > a$ the hyperbola. At this point it is easy to see what is going on. Condition (14) defines an ellipse, and it is clear that the sum of the distances of the point z from the points f_1 and f_2 cannot be less than the distance between the points f_1 and f_2. Thus to obtain a genuine ellipse we must have $2a > 2f$. For the hyperbola, the difference between $l(z, f_1)$ and $l(z, f_2)$ cannot be greater than $2f$, and so for a genuine hyperbola, we must have $2a < 2f$.

The equations (25) and (26) make it easy for us to see the form of ellipses and hyperbolas.

The ellipse is oval shaped and it is symmetrically located with respect both to the axis of abscissas and the axis of ordinates. (The reader may wish to refer back to Fig. 20.) The ellipse intersects the axis of abscissas in the two points $(a, 0)$ and $(-a, 0)$. It intersects the axis of ordinates in the two

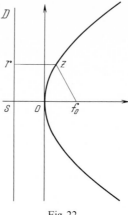

Fig. 22

points $(0, b)$ and $(0, -b)$. The inequality $a \geq b$ holds for all ellipses. If $a = b$, the two foci coincide and the ellipse is a circle.

The hyperbola defined by equation (26) is also symmetric with respect to both axes: the reader may refer to Fig. 21. It intersects the axis of abscissas in the two points $(a, 0)$ and $(-a, 0)$. It does not intersect the axis of ordinates. Thus the hyperbola consists of two parts, one to the right of the axis of ordinates and the other to the left. This parts are called the *two branches of the hyperbola.*

We will now give a geometric description of a *parabola* and derive an equation for it. A parabola is defined by a single *focus*, which we denote by f_0 and by a certain straight line D, which we call *the directrix*. Given a point z in our plane, we drop a perpendicular zr from z to the directrix D, as in Fig. 22. Let K denote our parabola. The curve K is defined as the set of all points z for which the length of the line segment zr is equal to the distance from z to the focus f_0. To derive an equation for K, we choose our coordinate axes as follows. The axis of abscissas is the line going through f_0 and perpendicular to the directrix D, with the direction running from s (the point where the axis intersects D) to f_0. For the origin we choose the midpoint of the line segment sf_0, so that the axis of ordinates is as sketched in Fig. 22. Let f denote the length of the line segment sf_0. For any point $z = (x, y)$ in the plane, the distance from z to the directrix D is equal to

$$l(z, r) = |\tfrac{1}{2}f + x|.$$

The distance from z to the focus f_0 is equal to

$$l(z, f_0) = +\sqrt{(x - \tfrac{1}{2}f)^2 + y^2}.$$

Therefore the parabola K is described by the equation

$$|\tfrac{1}{2}f + x| = +\sqrt{(x - \tfrac{1}{2}f)^2 + y^2}.$$

Square both sides of this equation and cancel terms from both sides to obtain

$$2fx = y^2. \tag{27}$$

This is an *equation for the parabola K*. Let us write $2f = \dfrac{1}{\alpha}$. The equation (27) then is

$$x = \alpha y^2.$$

Thus the form of the parabola is as described in equation (6) and sketched in Fig. 14: only x and y are interchanged.

We will now make some general observations about the equations for curves that we have just derived. For the sake of definiteness, we will consider the equation (13) for a circle with center at the origin: our remarks apply equally to all of the curves studied. We rewrite (13) as

$$x^2 + y^2 - r^2 = 0. \tag{28}$$

We next write

$$F(x, y) = x^2 + y^2 - r^2. \tag{29}$$

The right side of this equality is a specific polynomial and the left side is the expression $F(x, y)$. Just as in equation (1), we can regard the expression $F(x, y)$ as an abbreviation for the polynomial on the right side of (29). We can therefore write (28) as

$$F(x, y) = 0. \tag{30}$$

Now, in the right side of (29), we may replace x and y by any numbers we like, and obtain a specific number. That is, $F(x, y)$ is a function of the two independent variables x and y. This leads us to the notion of a function of two variables. We say that $F(x, y)$ is a *function of the two independent variables x and y* if, whenever x and y are given numerical values, we have a fixed numerical value for $F(x, y)$. The equality

$$z = F(x, y)$$

defines z as a dependent variable, that is, as a function of the two independent variables x and y. We may now say that if $F(x, y)$ is a function of the two independent variables x and y, then the equation

$$F(x, y) = 0 \tag{31}$$

defines a certain curve in the plane P. This curve consists exactly of the points (x, y) whose coordinates x and y satisfy the condition (31).

Suppose that $F(x, y)$ and $G(x, y)$ are two functions of the independent variables x and y. We may ask for solutions of the two equations

$$\left.\begin{array}{l} F(x, y)=0, \\ G(x, y)=0, \end{array}\right\} \tag{32}$$

in the variables x and y. Each equation separately,

$$F(x, y)=0 \tag{33}$$

and

$$G(x, y)=0, \tag{34}$$

defines a certain curve in the plane P. Plainly a pair of numbers (x, y) satisfies both of the equations (32) if and only if the point (x, y) lies on both of the curves defined by (33) and (34). That is, solving the simultaneous equations (32) is the same as finding the points of intersection of the two curves (33) and (34).

Suppose that the equation (30) can be solved explicitly for y in terms of x, that is, suppose that (30) is equivalent to a relation $y = f(x)$. Then (30) is the same description of a curve as we gave originally in formula (2). However, in solving equations of the form (30) we frequently encounter complications. We illustrate this with the equation (28) for a circle. If we solve for y in terms of x, we obtain

$$y = \sqrt{r^2 - x^2}. \tag{35}$$

The right side of (35) has meaning only for x such that $|x| \leq r$. That is, the function on the right side of (35) is not defined for all values of x. Furthermore, the square root of a positive number has two values, one positive and one negative. Therefore the equality (35) is actually two equalities:

$$y = +\sqrt{r^2 - x^2}. \tag{36}$$

and

$$y = -\sqrt{r^2 - x^2}. \tag{37}$$

The equation (36) defines a semicircle lying above the axis of abscissas, while (37) defines a semicircle lying below this axis. Thus the circle defined by (28) alone is described also by the two equations (36) and (37). The single equation (28) is a more natural way to describe our circle than are the two equations (36) and (37).

In conclusion we point out that equations of the form (30) do not always define curves in the plane. For example, consider the equation

$$x^2 + y^2 + r^2 = 0. \tag{38}$$

If r is a positive number, there are no pairs of numbers (x, y) that satisfy (38). That is, (38) in this case defines no curve whatever. If $r = 0$, then there is exactly one point, namely $(0, 0)$, that satisfies (38). It would be unnatural to consider a single point as a curve. Nevertheless, one sometimes regards the equation

$$x^2 + y^2 = 0$$

as defining a circle of zero radius, consisting of the point $(0, 0)$ alone.

Example 2. Consider an ellipse, an hyperbola, or a parabola, with equations (25), (26), and (27), respectively. Let us subject these curves to a dilation about the origin. This means that we move each point $z = (x, y)$ in the plane to the point $z_1 = \alpha z$, the positive number α being called *the coefficient of dilation*. Writing $z_1 = (x_1, y_1)$, we plainly have

$$x = \frac{x_1}{\alpha} \quad \text{and} \quad y = \frac{y_1}{\alpha}. \tag{39}$$

To find equations for the image curves of our ellipse, hyperbola, and parabola when we subject them a dilation, we substitute the values for x and y in (39) into the equations (25)–(27). This yields for the ellipse

$$\frac{x_1^2}{\alpha^2 a^2} + \frac{y_1^2}{\alpha^2 b^2} = 1;$$

for the hyperbola

$$\frac{x_1^2}{\alpha^2 a^2} - \frac{y_1^2}{\alpha^2 b^2} = 1;$$

and for the parabola

$$\frac{y_1^2}{\alpha^2} = 2f \frac{x_1}{\alpha}.$$

These equations have the form

$$\frac{x_1^2}{a_1^2} + \frac{y_1^2}{b_1^2} = 1, \quad \frac{x_1^2}{a_1^2} - \frac{y_1^2}{b_1^2} = 1, \quad y_1^2 = 2f_1 x_1,$$

where

$$a_1 = \alpha a \quad \text{and} \quad b_1 = \alpha b$$

for the ellipse and hyperbola and

$$f_1 = \alpha f$$

for the parabola. We have thus proved that ellipses, hyperbolas, and parabolas located with respect to the coordinate axes as we have placed them above go into the same types of curves under dilations, and (for ellipses and hyperbolas) we have

$$\frac{b_1}{a_1} = \frac{b}{a}. \tag{40}$$

Suppose that we have two ellipses or two hyperbolas defined by numbers a and b and a_1 and b_1, respectively. If the equality (40) holds, then the ellipses (hyperbolas) are obtained from each other by a dilation. Note too that every parabola can be obtained from every other parabola by a dilation. (All of this applies only to ellipses, hyperbolas, and parabolas in our standard position.)

Example 3. We present here a unified definition of ellipses, hyperbolas, and parabolas, similar to our definition of the parabola. Consider a point f_0 in the plane, which we call *the focus*, and any line D in the plane not containing f_0, which we call *the directrix*. We also specify some positive number k. We define a curve K in the plane as the set of all points z for which k times the distance of z from D is equal to the distance from z to f_0. That is, K is the set of all points z such that

$$k \, l(z, r) = l(z, f_0), \tag{41}$$

where r is the foot of the perpendicular from z to the line D. We already know that the choice $k=1$ gives us a parabola. It turns out that for $k<1$, the curve K is an ellipse, and for $k>1$, the curve K is an hyperbola. Let us prove this.

We choose our system of coordinates as follows. First draw a line through f_0 perpendicular to the directrix D. This will be our axis of abscissas. Let s be the point where the axis of abscissas meets D: see Fig. 23. The direction on the axis of abscissas will be chosen differently for the two cases $k>1$ and $k<1$. For $k>1$, we choose the direction as running from s to f_0. For $k<1$, we choose the direction as running from f_0 to s. (The reader may find it helpful to make sketches for the two cases: the case $k<1$ is illustrated in Fig. 23.) If we make our sketch so that the axis of abscissas is horizontal and goes from left to right, then for $k>1$, s lies to the left of f_0; for $k<1$, f_0 lies to the left of s. Since $k \neq 1$, there is a positive number a such that

$$l(f_0, s) = \left| ak - \frac{a}{k} \right|. \tag{42}$$

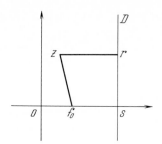

Fig. 23

Now start from the point f_0 and lay off an interval in the negative direction along the axis of abscissas with length ak. Let o be the left endpoint of this interval. A short calculation based on (42) shows that both f_0 and s lie to the right of o. Furthermore, both for $k>1$ and $k<1$, f_0 is at distance ak from o and s is at distance $\dfrac{a}{k}$ from o. We take the point o as our origin of coordinates. Thus the directrix D is parallel to the axis of ordinates and goes through the point s, whose coordinates are $\left(\dfrac{a}{k},0\right)$. Given a point $z=(x,y)$ in the plane, we have

$$l(z,r)=\left|x-\frac{a}{k}\right|$$

and

$$l(z,f_0)=+\sqrt{(x-ak)^2+y^2}.$$

Substitute these values in equation (41) and square both sides. We obtain

$$k^2\left(x-\frac{a}{k}\right)^2=(x-ak)^2+y^2.$$

Simple manipulations give

$$x^2(1-k^2)+y^2=a^2(1-k^2),$$

and so

$$\frac{x^2}{a^2}+\frac{y^2}{a^2(1-k^2)}=1.$$

Finally write f for the number ak, obtaining

$$\frac{x^2}{a^2}+\frac{y^2}{a^2-f^2}=1.$$

Thus an equation of our curve K has exactly the form (24). For $k<1$, $a-f^2$ is positive, and so the cuve K is an ellipse. For $k>1$, a^2-f^2 is negative, and so the curve K is an hyperbola. We have thus obtained a new geometric property of the ellipse and the hyperbola, *viz.*, the existence of the directrix D and the description (41) of the ellipse and hyperbola. It is of interest to note that we used the method of coordinates in our proof, although coordinates have nothing to do with our final result.

It is easy to see that two ellipses (hyperbolas) with the same value of k can be obtained from each other by a dilation. (Compare this with Example 2.)

Example 4. In elementary geometry, much attention is lavished on problems of constructions using only compass and straightedge. It is known that it is impossible using only compass and straightedge to trisect angles. That is, one cannot divide a general angle into three equal subangles making use only of compass and straightedge. However, if we allow ourselves to use hyperbolas with $k=2$ (as in Example 3), it is possible to trisect all angles. Let us carry out the construction.

In a plane P, choose any Cartesian system of coordinates, as in §1. Let K' be a circle of arbitrary positive radius with center at the origin o. Let e be the point of intersection of K' with the positive semi-axis of abscissas. Let f_0 be any point on the circle K' that lies in the first quadrant: see Fig. 24. We will trisect the angle eof_0. Let α and β be the points on the arc ef_0 of the circle K' for which the angles $eo\alpha$, $\alpha o\beta$, and βof_0 are all equal. Drop a perpendicular αr from the point α to the axis of abscissas. It is clear that the line segment αf_0 is twice as long as the line segment αr, that is, we have

$$2l(\alpha, r) = l(\alpha, f_0). \tag{43}$$

We construct the hyperbola K with $k=2$ and having f_0 for its focus and the axis of abscissas as its directrix. Condition (43) guarantees that the point α

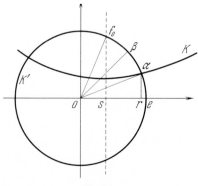

Fig. 24

lies on the hyperbola K (again see Fig. 24). Thus we trisect the angle $e\,o\,f_0$ by finding the point of intersection of the circle K' and the hyperbola K. For this it suffices to have one template of an hyperbola such that $k=2$. If the template is constructed so that the distance from the focus to the directrix is γ, then we must draw the circle K' with radius so that the ordinate of the point f_0 is equal to γ.

§6. Parametric Representation of Curves

Up to this point, we have described curves in the plane by a single equation connecting x and y. There is another way to describe curves, which is called *the parametric method*. This method offers certain advantages, as we shall see. We describe the curve directly as the path traced out by a point moving in the plane. Specifically, we consider two equations

$$\left.\begin{array}{l} x=x(t), \\ y=y(t), \end{array}\right\} \tag{44}$$

where $x(t)$ and $y(t)$ are functions of the same independent variable t. As t varies on the line (increasing, say), the point

$$z(t)=(x(t), y(t))$$

moves in the plane, and traces out a certain curve K. Since we are considering the movement of a point in the plane, it is natural to suppose that the point $z(t)$ does not jump suddenly from one position to another (see Example 1 in §4). For the moment we will not try to make this notion precise, but take it as intuitively clear. Very frequently the number t, which is called a *parameter*, may be thought of as time. This is the case when we describe some physical process that takes place in time, and x and y are physical quantities that vary with time. For example, we may consider a piece of electrical apparatus, a network perhaps, and let x be the voltage drop as current runs through the network, while y may be the number of amperes in the current flow. Also parametric representation can be very convenient in purely geometric considerations. We illustrate this with the example of a straight line.

In giving parametric equations for a straight line, let us make use of vector notation. Consider a plane P with a system of Cartesian coordinates, as sketched in Fig. 25. Let M be a straight line in the plane P. Let z_0 be an arbitrary point on the line M. Recall that we use the letter z_0 also to denote the vector $o\,z_0$. Let e be the vector of length 1 that goes through the origin and is parallel to M. There are two choices for this vector, one being the negative of the other. Choose either one: our construction does not depend

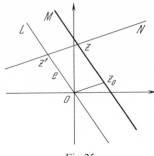

Fig. 25

upon which we select. It turns out that the line M can be described in the vector form

$$z = z_0 + te. \tag{45}$$

This means that the point z (equivalently, the terminal point of the vector z) runs through the entire line M as t runs through all real values.

Let us prove this. First draw a line L parallel to M and passing through the origin. The vector e lies on L, and as was remarked in formula (6) of Chapter I, every point z' lying on L has the form

$$z' = te$$

for some real number t. If the line M happens to go through the origin, then L and M coincide. The two points z_0 and te that appear on the right side of (45) are both on L. It is clear that beginning with an arbitrary point z_0 of L and laying off intervals te of arbitrary length and direction from z_0, we obtain all the points of the line L.

Suppose now that the lines L and M are distinct. Draw a line N parallel to the vector z_0. Let z' and z be the points of intersection of N with L and M respectively. We sketch the construction in Fig. 25. Since $oz'zz_0$ is a parallelogram, we see that

$$z = z_0 + z'. \tag{46}$$

Now suppose that z is an arbitrary point of the line M. We draw our line N through this point z. There is a number t such that $z' = te$. From (46), we see that the point z can be written in the form (45). If t is an arbitrary number and $z' = te$, we draw our line N through this point z'. Again (46) shows that the point z defined by formula (45) lies on the line M.

From the vectorial parametric equation (45) for a straight line, we can easily obtain the usual coordinate equation. The vector e has unit length, and so we have

$$e = (\cos \alpha, \sin \alpha),$$

where α is the angle between the positive semi-axis of abscissas and the vector e (measure this angle in the counterclockwise direction). Writing $z_0 = (x_0, y_0)$ and $z = (x, y)$, we rewrite the vector equation (45) in the form of two coordinate equations

$$x = x_0 + t \cos \alpha,$$
$$y = y_0 + t \sin \alpha.$$

We eliminate the parameter t in these equations by multiplying the first by $\sin \alpha$, the second by $\cos \alpha$, and subtracting the second from the first. We then obtain

$$(y - y_0) \cos \alpha = (x - x_0) \sin \alpha. \tag{47}$$

If the line M is not vertical, the cosine of α is different from zero. In this case, we divide both sides of (47) by $\cos \alpha$ to obtain

$$y - y_0 = (x - x_0) \tan \alpha.$$

Since $\tan(\alpha + \pi) = \tan \alpha$, the last equation does not depend upon our choice of the direction of the vector e parallel to the line M. This equation is frequently written in the form

$$y - y_0 = k(x - x_0). \tag{48}$$

The number α is called *the angle of inclination of the line M*, and the number $k = \tan \alpha$ is called *the slope of M*. The point $z_0 = (x_0, y_0)$ is any point whatever on the line M.

The equation (48) suffers from the drawback that it cannot be used to represent vertical lines. If M is vertical, we have $\cos \alpha = 0$ and $\sin \alpha = 1$ (if e is taken as the vector $(0, 1)$) and so equation (47) has the form

$$x = x_0 \tag{49}$$

for a vertical line containing the point (x_0, y_0).

We next present parametric equations for an ellipse. They are:

$$\left. \begin{array}{l} x = x(t) = a \cos t, \\ y = y(t) = b \sin t. \end{array} \right\} \tag{50}$$

From these parameteric equations, we obtain the usual equation (25) by eliminating the parameter t. To do this, divide the first equation (50) by a and the second by b. Then square both equations and add them together. This yields equation (25). The parametric equations (50) not only yield an ellipse as a geometric figure but also describe it as the path traced out by a moving point. In fact, as t runs from 0 to 2π, the point (50) traces out our ellipse in the counterclockwise direction, beginning at $(a, 0)$ and returning to $(a, 0)$ when t has reached the value 2π. For this reason, the ellipse is called *a closed curve*. It "closes up" after one circuit of the entire curve.

It is also easy to set down parametric equations for a circle of radius r and center at the point $z_0 = (x_0, y_0)$. These equations may be taken as

$$\left.\begin{aligned} x = x(t) = x_0 + r\cos t, \\ y = y(t) = y_0 + r\sin t. \end{aligned}\right\} \tag{51}$$

The usual equation for a circle can quickly be obtained from the parametric equations (51). Move x_0 and y_0 to the left sides, square, and add. This gives us equation (12). As t runs from 0 to 2π, the point $z(t) = (x(t), y(t))$ traces out the entire circle in the counterclockwise direction. Note that $z(0) = z(2\pi)$. In the degenerate case $r = 0$, the point $z(t)$ does not move as t varies, but simply stays put at (x_0, y_0). All the same, it is sometimes convenient to consider this trivial case as a closed curve as well. The important point is that $z(0)$ be equal to $z(2\pi)$.

Example 5. Let us find an equation for a line M passing through two distinct points $z_0 = (x_0, y_0)$ and $z_1 = (x_1, y_1)$. If $x_0 = x_1$, the line M is vertical and we can write its equation as (49). If $x_0 \neq x_1$, we determine the number k in equation (48) by putting z_1 for z. This gives us

$$y_1 - y_0 = k(x_1 - x_0),$$

so that

$$k = \frac{y_1 - y_0}{x_1 - x_0}. \tag{52}$$

That is, the slope of the line going through the points z_0 and z_1 is defined by (52).

§7. Closed Curves

We now consider an arbitrary closed curve K having parametric equations (44)

$$x = x(t),$$

$$y = y(t).$$

We suppose that the parameter t runs from 0 to 2π. The property of being closed is expressed by the equalities $x(0) = x(2\pi)$, $y(0) = y(2\pi)$. That is, the moving point $z(t) = (x(t), y(t))$ returns to its starting point:

$$z(2\pi) = z(0). \tag{53}$$

We do not exclude the possibility that $z(t_1)=z(t_2)$ for distinct values of the parameter t besides 0 and 2π. This phenomenon is called *a self-intersection of the curve K*. In order for K to be closed, we need only that (53) hold.

We suppose also that the closed curve K does not pass through the origin. For such a curve, we can define an integer j called *the winding number of K about the origin*. The number j shows how many times the point $z(t)$ "goes around" the origin as t increases from 0 to 2π.

To give a formal definition of the number j, we introduce polar coordinates $\rho(t)$ and $\varphi(t)$ for the point $z(t)$, as in formula (13) of Chapter I. The radius $\rho(t)$ of the point $z(t)$ is of course uniquely determined by the point $z(t)$. Since $z(t)$ moves in the plane without any breaks, the function $\rho(t)$ is continuous, which is to say that it too varies with t exhibiting no breaks or jumps. The angle $\varphi(t)$ on the other hand is not uniquely defined by $z(t)$: it is determined only up to an additive constant $2k\pi$, where k is an integer. Nevertheless, if we choose some fixed angle $\varphi(0)$ for the initial point $z(0)$ of our curve, then the function $\varphi(t)$ is defined uniquely for all t, $0\leq t\leq 2\pi$, if we require in addition that it vary continuously with t. Note that adding a nonzero number $2k\pi$ to (t) at some point t would introduce a jump in the function $\varphi(t)$. Since K does not pass through the origin, there is no value of the parameter t where the angle $\varphi(t)$ is undefined. The equality (53) in polar coordinates becomes

$$[\rho(2\pi), \varphi(2\pi)]=[\rho(0), \varphi(0)].$$

Since ρ is uniquely defined, the last equality means that $\rho(2\pi)=\rho(0)$. However, this need not be the case for the function $\varphi(t)$. Instead, we have

$$\varphi(2\pi)-\varphi(0)=2j\pi, \tag{54}$$

where j is some integer. This integer j is *the winding number of the curve K about the origin*. The integer j can be defined only for closed curves that do not pass through the origin. (In Example 7 we will see what happens when K passes through the origin.)

We will now show that the winding number j of a curve K does not change if K is continuously deformed in the plane P without passing through the origin during this deformation.

To get an intuitive idea of a continuous deformation of a curve K, let us think of K as made out of some physical material, such as a piece of thread tied together at its ends. A continuous deformation of K is any motion of the thread in the plane, where we are allowed to bend, stretch, and shrink the thread as much as we like.

Mathematically, a continuous deformation of a closed curve K is defined as follows. We introduce a new parameter s, varying between limits a and b: $a\leq s\leq b$. For every value of s, there is a closed curve, which we denote by $K(s)$. Thus $K(s)$ is a function of the variable s. The values of the function $K(s)$ are neither numbers nor vectors, but entire closed curves. For a fixed value of the parameter s, the closed curve $K(s)$ is the path traced out by the point

$$z(t, s) = [\rho(t, s), \varphi(t, s)]$$

as t runs from 0 to 2π. Note also that

$$z(2\pi, s) = z(0, s).$$

We suppose that the closed curve $K(s)$ varies continuously with s and that no curve $K(s)$ passes through the origin. Thus the winding number about the origin is defined for every curve $K(s)$. We have

$$\varphi(2\pi, s) - \varphi(0, s) = 2j(s)\pi \tag{55}$$

for every value of the parameter s. Since the curve is deformed continuously, the functions $\varphi(2\pi, s)$ and $\varphi(0, s)$ vary with s without exhibiting any jumps. Therefore $j(s)$ also varies without any jumps as s varies. Since the number $j(s)$ is an integer, it must be constant if it is to vary without any jumps.

Therefore all of the curves $K(s)$ have the same winding number $j(s)$. This fact is highly important for an application that we will give in the following section.

Example 6. We will compute the winding number of a circle K having parametric equations (51). We suppose that K does not pass through the origin. There are two possibilities: 1) the origin does not lie in the interior of K; 2) the origin lies in the interior of K.

In the first case, we can draw exactly two tangent lines from the origin 0 to the circle K. Designate them by $o\,a_1$ and $o\,a_2$, where a_1 and a_2 are the points of tangency on the circle K. The geometrical situation is sketched in Fig. 26. We suppose that a_1 and a_2 are labelled in such a way that $o a_2$ is obtained from $o a_1$ by a counterclockwise rotation through an angle α such that $0 < \alpha < \pi$. Let φ_1 be any angle for the point a_1. Then the number $\varphi_1 + \alpha = \varphi_2$ is an angle for the point a_2. We can choose an angle $\varphi(0)$ for the

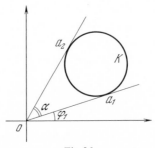

Fig. 26

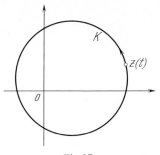

Fig. 27

initial point $z(0)$ in such a way that $\varphi_1 < \varphi(0) < \varphi_2$. During its entire motion, the point $z(t)$ remains within the sector $a_1 \, o \, a_2$. Consequently the angle $\varphi(t)$ satisfies the inequalities

$$\varphi_1 \leqq \varphi(t) \leqq \varphi_2$$

for all values of t, $0 \leqq t \leqq 2\pi$. Since $\alpha < \pi$, the preceding inequalities show that

$$\varphi(2\pi) - \varphi(0) \leqq \alpha < \pi.$$

Hence the integer j (which is obviously in this case nonnegative) is less than $\frac{1}{2}$ and so must be 0.

We sketch case 2 in Fig. 27. In this case, it is geometrically obvious that the vector $z(t)$ moves counterclockwise around the circle as t increases from 0 to 2π. Hence the angle $\varphi(t)$ increases from its initial value $\varphi(0)$ to $\varphi(0) + 2\pi$. Therefore j is equal to $+1$ for this curve.

Example 7. Let us see what happens to the winding number of the circle K as the circle K moves in the plane from a position in case 1 to a position in case 2 (see Example 6). Let $K(s)$ be the circle of radius 1 and center on the axis of abscissas at the point $(1+s, 0)$, where s is a small number. For $s > 0$, the circle $K(s)$ does not contain the origin in its interior. Hence its winding number about the origin is zero (see Fig. 28). For $s < 0$, the origin lies in the interior of $K(s)$ and so its winding number about the origin is equal to $+1$ (see Fig. 29).

For $s > -2$, the initial point $z(0) = (2+s, 0)$ of the circle $K(s)$ lies on the positive semi-axis of abscissas. We can thus take the angle $\varphi(0)$ to be 0. With this choice, we will study the continuous function $\varphi(t)$, the angle of the point $z(t)$, as $z(t)$ traces out the circle $K(s)$. We will do this first for small positive values of s and then for negative values of s with small absolute values.

For s small and positive, we draw the two tangents $o a_1$ and $o a_2$ to the circle $K(s)$ as sketched in Fig. 28. We can write an angle of the point a_1 as $-\frac{1}{2}\alpha$ and an angle of the point a_2 as $\frac{1}{2}\alpha$, where α is a number such that

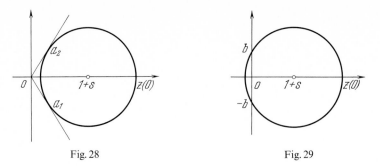

Fig. 28   Fig. 29

$0 < \alpha < \pi$. For sufficiently small s, α will be as close as you wish to π. The point $z(0)$ is the point of $K(s)$ that is the farthest to the right. Now consider the point $z(t)$ moving on the circle $K(s)$ with increasing t. Plainly the angle $\varphi(t)$ increases up to the point where $z(t)$ reaches the point a_2. At this point $\varphi(t)$ attains its largest value, namely $\frac{1}{2}\alpha$. For small enough s, this value will be arbitrarily close to $\frac{1}{2}\pi$. As t continues to increase, the angle $\varphi(t)$ steadily decreases, assuming the value 0 for $t = \pi$ and then taking on negative values, decreasing until it reaches its minimum value $-\frac{1}{2}\alpha$ at the value of t that places $z(t)$ at the point a_1. As the point $z(t)$ moves from a_1 back to $z(2\pi)$ $= (2+s, 0)$, the angle $\varphi(t)$ increases from $-\frac{1}{2}\alpha$ to 0. We have plotted the graph of a sample function $\varphi(t)$ (s is small) in Fig. 30.

We now consider $K(s)$ for negative s. If $-2 < s < 0$, the circle $K(s)$ intersects the axis of ordinates in two distinct points $(0, b)$ and $(0, -b)$, where b is a positive number that is arbitrarily close to 0 if $|s|$ is sufficiently small. Let the point $z(t)$ trace out the circle $K(s)$ as t increases from 0 to 2π. The angle $\varphi(t)$ of the point $z(t)$ increases steadily with t. Its rate of increase is very rapid as t increases from the point where $z(t) = (0, b))$ to the point where $z(t) = (0, -b)$. In this interval, $\varphi(t)$ increases from $\frac{1}{2}\pi$ to $\frac{3}{2}\pi$. From this value of t up to 2π, $\varphi(t)$ continues to increase, but much more slowly. In Fig. 31, we have plotted the graph of $\varphi(t)$ for a sample negative s with small absolute value.

For $s = 0$, we get the circle $K(0)$, which passes through the origin. See the sketch in Fig. 32. A famous theorem of elementary geometry shows that

$$\varphi(t) = \tfrac{1}{2}t$$

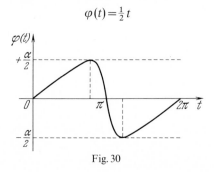

Fig. 30

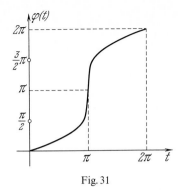

Fig. 31

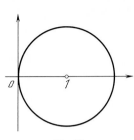

Fig. 32

for $0 \leq t < \pi$. At $t = \pi$, the point $z(t)$ is the origin, so that its angle is undefined. After $z(t)$ has passed through the origin, we can consider that its angle is either close to $-\frac{1}{2}\pi$ or to $\frac{3}{2}\pi$. The graph of the function $\varphi(t)$ has a jump at the point $t = \pi$. We can suppose that this jump has either of two forms. One form is sketched in Fig. 33, with $\varphi(t)$ close to $-\frac{1}{2}\pi$ for t larger than π and close to π. The other form is sketched in Fig. 34, with $\varphi(t)$ close to $\frac{3}{2}\pi$ for t larger than π and close to π.

Now let s go to zero through positive values. The graph of the function $\varphi(t)$ converges to the graph in Fig. 33. As s goes to 0 through negative values, the graph of $\varphi(t)$ converges to the graph in Fig. 34.

This is the mechanism of discontinuous change of the winding number $j(s)$ defined by (55) when the curve $K(s)$ passes through the origin for some value of the parameter s.

Example 8. Let n be a nonzero integer, positive or negative. Define a curve K_n with parametric polar equations

$$\left.\begin{array}{l} \rho = \rho(t) = 1 + s \cos t, \\ \varphi = \varphi(t) = nt. \end{array}\right\} \tag{56}$$

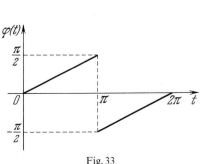

Fig. 33

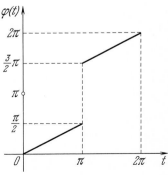

Fig. 34

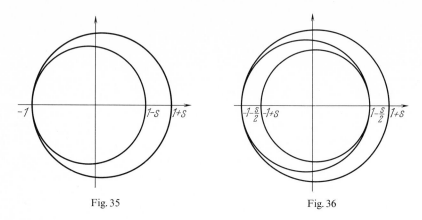

Fig. 35   Fig. 36

The number s is a constant such that $0 \leq s < 1$, and the parameter t runs from 0 to 2π. The restrictions on s guarantee that $\rho(t)$ is positive for all t, so that K_n does not pass through the origin. The angle $\varphi(t)$ as specified in (56) is a continuous function of t, and it is obvious that $\varphi(0)=0$, $\varphi(2\pi)=2n\pi$. The winding number, as (54) shows, is equal to n. Thus the winding number of a closed curve can be any nonzero integer. We sketch the curve K_2 in Fig. 35 and the curve K_3 in Fig. 36. Note that K_2 has one point of self-intersection and that K_3 has two. Both sketches are for the case $s>0$.

For $s=0$, the point $z(t)=[\rho(t), \varphi(t)]$ moves on the circle K' with radius 1 and center at the origin. The angle $\varphi(t)$ increases n times faster than t (for positive n), and so as t runs from 0 to 2π, the point $z(t)$ goes around the circle n times. For negative n, it goes around the circle $-n$ times, but in the clockwise direction.

§8. Polynomials in a Complex Variable

By analogy with formula (1) of §4, we write the equality

$$f(z)=a_0 z^n+a_1 z^{n-1}+ \cdots +a_n, \tag{57}$$

where the numbers $a_0, a_1, \ldots, a_n$ are now any complex constants. The right side of (57) is a polynomial with complex coefficients in the variable z. The expression $f(z)$ on the left side of (57) is an abbreviation for the polynomial on the right side. Since we can carry out algebraic operations on complex numbers, we can substitute any fixed complex number in the right side of (57) and get a fixed complex number out. That is, $f(z)$ is a *complex function of the complex variable z*.

A complex function of a complex variable z need not be a polynomial in z. Any rule that assigns a fixed complex value to every complex number is in fact such a function: we write $f(z)$ for such a function or rule. Given $f(z)$, we write

$$w = f(z) \tag{58}$$

and obtain *a dependent complex variable w* defined as the function $f(z)$ *of the complex independent variable* or *argument z.* In this section we will center our attention on polynomial functions defined as in (57).

For real polynomials $f(x)$ in a real variable x, we have a simple geometric method of picturing the function $f(x)$: we simply graph it, as explained in §4. No such simple device is available for representing complex polynomials in a complex variable z. Nevertheless, there is a geometric approach to the description of complex functions of a complex variable. We think of two complex planes, one called the Z plane, where we find the complex variable z. The other is called the W plane, in which we find the dependent variable w (notation as in (58)). We think of the function $f(z)$ as mapping the Z plane into the W plane, and this point of view opens up great possibilities for studying the function $f(z)$. We will use this point of view to prove the fundamental theorem of algebra.

We say that the polynomial in (57) has degree n if the coefficient a_0 of the power z^n is different from zero. To make this crystal clear, we will consider polynomials of the form

$$f(z) = z^n + a_1 z^{n-1} + \cdots + a_n. \tag{59}$$

The first (and essential) step in proving the fundamental theorem of algebra is to show that every polynomial $f(z)$ of the form (59) and with $n \geq 1$ admits at least one root, that is, that there exists a complex number γ_1 such that

$$f(\gamma_1) = 0.$$

The geometric idea of our proof of this fact is as follows. We write the complex number z in trigonometric form, that is,

$$z = r[\cos t + i \sin t]. \tag{60}$$

Here r is the modulus of the complex number z and t is its argument. Let us fix the positive number r and let t vary from 0 to 2π. As this happens, the point z describes in the plane Z the circle of radius r and center at the complex number 0. We denote this circle by $K(r)$, to emphasize its dependence on the choice of the number r. The circle $K(r)$, as t runs from 0 to 2π, is a closed curve, since for $t = 0$ and for $t = 2\pi$, we obtain the same complex number $z = r$.

We now consider the curve in the W plane described by the point $w = f(z)$ (defined in (58)) as z describes the circle $K(r)$. We may write

$$w(t) = f(r(\cos t + i \sin t)).$$

We write $w(t)$ since we have held r fixed and allow only t to vary. It is clear that

$$w(0) = w(2\pi) = f(r).$$

Hence the curve described by $w(t)$ as t runs from 0 to 2π is closed. We denote this curve by $L(r)$, again emphasizing its dependence upon r. If $L(r)$ does not pass through the origin, we can define its winding number about the origin, as was done in §7, p. 51. We will show that this number is equal to the degree n of the polynomial $f(z)$ for all sufficiently large values of r.

We begin by computing the winding number of $L(r)$ for the simplest polynomial of degree n, namely

$$w = f(z) = z^n. \tag{61}$$

We write w in trigonometric form:

$$w = \rho(\cos\varphi + i\sin\varphi).$$

For the polynomial (61), we find at once that

$$\rho = r^n \quad \text{and} \quad \varphi = nt.$$

As t runs from 0 to 2π, φ runs from 0 to $2n\pi$. Therefore the winding number of the curve $L(r)$ is equal to n for the polynomial (61).

To compute the winding number of the curve $L(r)$ for the general polynomial (59), we introduce a one-parameter family of polynomials $f(z, s)$ defined by

$$w = f(z, s) = z^n + s(a_1 z^{n-1} + \cdots + a_n). \tag{62}$$

The parameter s is subject to the conditions $0 \leq s \leq 1$. For $s = +1$, we get the general polynomial (59) and for $s = 0$, we get the special polynomial (61).

We single out the polynomial in parentheses on the right side of (62) for separate examination. Let us write

$$g(z) = a_1 z^{n-1} + \cdots + a_n.$$

We want to estimate the modulus of the quantity $g(z)$. Since we are interested only in large values of r, we may suppose at the outset that $r \geq 1$. As was pointed out in formula (28) of Chapter I, the modulus of a sum is less than or equal to the sum of the moduli. Hence for $r \geq 1$, we find

$$|g(z)| \leq r^{n-1}(|a_1| + |a_2| + \cdots + |a_n|),$$

or

$$|g(z)| \leq c r^{n-1}, \tag{63}$$

where we have written c for $|a_1|+\cdots+|a_n|$. We have thus found what is called in mathematics an estimate for the modulus of the quantity $|g(z)|$, valid for all z such that $r=|z|\geq 1$.

It follows from (62) that

$$f(z,s)+(-s\,g(z))=z^n.$$

From this equality we obtain

$$|f(z,s)|+s\,|g(z)|\geq r^n.$$

To this inequality we apply the estimate (63), obtaining

$$|f(z,s)|+c\,r^{n-1}\geq r^n,$$

or

$$|f(z,s)|\geq r^n-c\,r^{n-1}. \tag{64}$$

Since we are interested only in large values of r, we may suppose that $r>c$. Then the right side of (64) is positive, so that

$$|f(z,s)|>0$$

for all r greater than $\max(1,c)$. Therefore the curve $L(r,s)$ does not pass through the origin for any value of s, $0\leq s\leq 1$. Since the dependence of $L(r,s)$ on s is certainly continuous, we see as in §7 that the winding number of $L(r,s)$ does not vary as s varies from 0 to 1. For $s=0$, we have computed the winding number and found it to be equal to n. Therefore the curve $L(r,1)$, which is just $L(r)$, also has winding number n.

It is now easy to show that the polynomial $f(z)$ as in (59) has at least one complex root. If $a_n=0$, it is clear from (59) that $f(0)=0$, so that the polynomial $f(z)$ has at least one root in this case.

Suppose then that $a_n\neq 0$. We consider the changes that the curve $L(r)$ undergoes as r decreases from a large positive value r_0 to zero. (We could take $r_0=1+c$, for example, which would ensure that $L(r_0)$ has winding number equal to n.) It is plain that the curve $L(r)$ is deformed continuously, and that for $r=0$, it shrinks to the single point $a_n\neq 0$. Thus for $r=0$, the point $w(t)$ does not move in the complex W plane and is the fixed point a_n. Since $a_n\neq 0$, it has an argument, which does not vary with t. Therefore we have $\varphi(2\pi)-\varphi(0)=0$, which is to say that the winding number of the degenerate curve $L(0)$ is zero. We have also proved that the winding number of the curve $L(r_0)$ is n. It follows that in the process of deforming, there must be some positive number r_1 less than r_0 such that the curve $L(r_1)$ passes through the origin: otherwise the curves $L(r_0)$ and $L(0)$ would have the same winding numbers. Since the curve $L(r_1)$ passes through the origin, there is a number t_1 such that $0\leq t_1<2\pi$ for which $w(t_1)=0$, which is to say

that
$$w = f(r_1(\cos t_1 + i \sin t_1)) = 0,$$

or

$$\gamma_1 = r_1(\cos t_1 + i \sin t_1)$$

is a root of the polynomial $f(z)$.

It is now easy to show that a polynomial (59) of degree n admits exactly n complex roots. Although proofs are given in elementary algebra, I will give one here. Let γ be an arbitrary complex number. Divide the polynomial $f(z)$ as in (59) by the binomial $z - \gamma$. Carrying out the division, we write

$$f(z) = f_1(z)(z - \gamma) + b, \tag{65}$$

where $f_1(z)$ is a polynomial of degree $n - 1$ and b depends only on γ. We write $f_1(z)$ as

$$f_1(z) = z^{n-1} + b_1 z^{n-2} + \cdots + b_n.$$

Now, in (65), let us take for γ a root γ_1 of $f(z)$. We get

$$f(z) = f_1(z)(z - \gamma_1) + b.$$

Setting $z = \gamma_1$ in this identity, we find $b = 0$. That is, if γ_1 is a root of $f(z)$, then the binomial $z - \gamma_1$ divides the polynomial $f(z)$:

$$f(z) = f_1(z)(z - \gamma_1).$$

If the polynomial $f(z)$ has degree $n \geq 2$, then the polynomial $f_1(z)$ has degree $n - 1 \geq 1$. By what we have already proved, we can write

$$f_1(z) = f_2(z)(z - \gamma_2),$$

where γ_2 is a root of the polynomial $f_1(z)$ and therefore also of $f(z)$. Continuing this process, we obtain the following decomposition for the polynomial $f(z)$:

$$f(z) = (z - \gamma_1)(z - \gamma_2) \cdots (z - \gamma_n). \tag{66}$$

The numbers

$$\gamma_1, \gamma_2, \ldots, \gamma_n \tag{67}$$

are obviously roots of the polynomial $f(z)$. Plainly some of them may be equal to each other. Thus the assertion that a polynomial of degree n has n roots does not mean that the polynomial admits n distinct roots, but only that the polynomial can be represented in the form (66). If a given number γ_1 occurs in the sequence (67) exactly once, then it is called *a simple root of*

$f(z)$. If the number γ_1 occurs exactly k times in the sequence (67), where $k>1$, then γ_1 is called *a multiple root of multiplicity k*, again of the polynomial $f(z)$.

We proved at the end of §3 a fact about polynomials with real coefficients. Namely, if a complex number γ is a root of such a polynomial $f(z)$, then the complex conjugate $\bar{\gamma}$ is also a root of $f(z)$. It turns out that the roots γ and $\bar{\gamma}$ have the same multiplicity. Let us prove this. Let γ be a complex nonreal root. Then $\bar{\gamma}$ is also a root and is distinct from γ. Therefore we can write $f(z)$ in the form

$$f(z)=(z-\gamma)(z-\bar{\gamma})f_2(z)=g(z)f_2(z),$$

where

$$g(z)=(z-\gamma)(z-\bar{\gamma})=z^2-(\gamma+\bar{\gamma})z+\gamma\bar{\gamma}.$$

Note that $g(z)$ has real coefficients. That is, the real polynomial $f(z)$ is divisible by the real polynomial $g(z)$. If we carry out this division by the rules of elementary algebra, we see that $f_2(z)$ also has real coefficients. Therefore if γ is a root of $f_2(z)$, its complex conjugate $\bar{\gamma}$ is also a root. Continuing this process, we see that complex nonreal roots γ and $\bar{\gamma}$ of a polynomial with real coefficients must have the same multiplicity.

Example 9. We subject the simplest polynomial of degree n,

$$f(z)=z^n,$$

to closer scrutiny. That is, we will try to gain a clearer picture of how the mapping f of the Z plane onto the W plane behaves. Again we use polar coordinates in both planes:

$$z=r(\cos t+i\sin t), \tag{68}$$

$$w=\rho(\cos\varphi+i\sin\varphi). \tag{69}$$

We thus have

$$\rho=r^n \quad \text{and} \quad \varphi=nt.$$

Let us fix t and let r assume all nonnegative values. Then z sweeps out a rectilinear ray $A(t)$ in the Z plane, with angle of inclination t. This is clear from (68). The points w corresponding to this ray $A(t)$ sweep out a rectilinear ray $B(\varphi)$ in the W plane, with angle of inclination $\varphi=nt$. That is, for fixed t, the mapping f of the Z plane onto the W plane is simply a mapping of the ray $A(t)$ onto the ray $B(\varphi)$, given by the formula

$$\rho=r^n. \tag{70}$$

To gain a clearer picture of the mapping of $A(t)$ onto $B(\varphi)$, let us plot the graph of the function (70), laying off r on the axis of abscissas and ρ on the

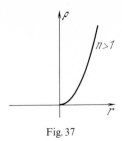

Fig. 37

axis of ordinates, as in Fig. 37. Note that $A(0)$ is the nonnegative semi-axis of abscissas in the Z plane, and that $B(0)$ is the nonnegative semi-axis of abscissas in the W plane. Now let t increase from 0 to $\dfrac{2\pi}{n}$. The rays $A(t)$ sweep out the angular region in the Z plane between the ray $A(0)$ and the ray $A\left(\dfrac{2\pi}{n}\right)$. The rays $B(\varphi)$ that correspond to the rays $A(t)$ carry out a complete rotation through an angle of 2π and sweep out the entire W plane. As t increases from $\dfrac{2\pi}{n}$ to $\dfrac{4\pi}{n}$, the rays $A(t)$ sweep out the angular region between $A\left(\dfrac{2\pi}{n}\right)$ and $A\left(\dfrac{4\pi}{n}\right)$. The corresponding rays $B(\varphi)$ sweep out the entire W plane once more. Thus as t increases from 0 to 2π, the rays $A(t)$ sweep out the entire Z plane exactly once. The corresponding rays $B(\varphi)$ complete n complete rotations and sweep out the W plane n times. That is, the mapping f carries the Z plane onto the W plane so that every point $w \neq 0$ is covered n times. The sole exception is the point 0 in the W plane, which is the image under f of only the point 0 in the Z plane. This mapping f is typical of the mappings that one encounters in the theory of functions of a complex variable.

Example 10. In proving the fundamental theorem of algebra, we considered the circle $K(r)$ in the Z plane with radius r and center at the complex number 0. The image of $K(r)$ under the mapping f defined in (59) is a closed curve $L(r)$ in the W plane. We proved that for sufficiently large r, say $r = r_0$, the winding number of the curve $L(r_0)$ is n. Under the hypothesis that a_n be different from 0, we shrank the curve $L(r)$ with decreasing r down to the single point $L(0)$. Since the winding number of $L(0)$ is 0 and the winding number of $L(r_0)$ is n, we conclude that the curve $L(r_1)$ must pass through the origin for some r_1 such that $0 < r_1 < r_0$. It would be natural to expect that the curve $L(r)$ must pass through the origin n times as r decreases from r_0 to 0. This would mean that the polynomial $f(z)$ admits n distinct roots. In point of fact, the polynomial $f(z)$ may have repeated roots. To explain what happens geometrically when repeated roots occur, we must admit the possibility that the curve $L(r)$ has little loops going around the origin in the W

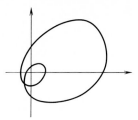

Fig. 38

plane, as sketched in Fig. 38. As r decreases, this loop may shrink to the point $w=0$. That is, by tightening the loop until it collapses to 0, one may find that $L(r)$ goes through the origin for only one value of r. This explains how roots can coincide with each other.

Supplement to Chapter II

We turn our attention to space once again. Suppose that we have a system of rectangular Cartesian coordinates in space, as constructed in the Supplement to Chapter I. We will consider the connections between algebraic relations and geometric figures that the establishment of a coordinate system makes possible.

1. Functions of Two Variables and their Graphs in Space. Let $f(x_1, x_2)$ be a function of the two variables x_1 and x_2. The equation

$$x_3 = f(x_1, x_2)$$

defines a surface S in space, consisting of all points of the form

$$x = (x_1, x_2, f(x_1, x_2)).$$

This surface may be regarded as the graph of the function $f(x_1, x_2)$ in space. Plainly it depends upon our choice of a coordinate system (see the corresponding discussion in §4 of the present chapter). Let x_1, x_2 be an arbitrary pair of numbers and let $p=(x_1, x_2)$ be the point in the $(1, 2)$ plane with coordinates x_1 and x_2. We designate the $(1, 2)$ plane by the symbol P. Now draw a line in space perpendicular to the plane P and passing through the point p. Lay off an interval on this line of length $|f(x_1, x_2)|$, and lying above the plane P if $f(x_1, x_2)$ is positive and lying below the plane P is $f(x_1, x_2)$ is negative. Our surface S consists of all of the endpoints x of these intervals, as x_1 and x_2 assume all real values.

If the function $f(x_1, x_2)$ is a constant, that is,

$$f(x_1, x_2) = c,$$

the surface S is the plane $P(c)$ parallel to the plane P and at distance $|c|$ from P. The plane $P(c)$ is above P if c is positive and is below P if c is negative. There is a natural system of Cartesian coordinates in the plane $P(c)$. For the first axis we take the intersection of $P(c)$ with the $(1, 3)$ plane and for the second axis we take the intersection of $P(c)$ with the $(2, 3)$ plane.

We consider next a more interesting function $f(x_1, x_2)$:

$$f(x_1, x_2) = \alpha(x_1^2 - x_2^2),$$

where α is a positive constant. To get a clear picture of the surface S defined by the corresponding equation

$$x_3 = \alpha(x_1^2 - x_2^2), \tag{71}$$

let us intersect this surface with planes $P(c)$. The surface S and the plane $P(c)$ intersect in a curve whose equation in $P(c)$ is

$$x_1^2 - x_2^2 = \frac{c}{\alpha}. \tag{72}$$

For $c = 0$, $P(c)$ is our coordinate plane P, and the equation (72) defines two straight lines. These are the lines bisecting the four quadrants. They have the equations $x_2 = \pm x_1$. For positive c, the curve of intersection is an hyperbola. Its branches lie to the right and the left of the second axis in $P(c)$. For negative c, the curve of intersection is also an hyperbola. In this case, its branches lie below and above the first axis in $P(c)$. Considering this description a bit further, we see that in the neighborhood of the origin, the surface S looks rather like a saddle: see the sketch in Fig. 39.

2. Functions of Three Variables and the Surfaces that Correspond to them. Let $F(x_1, x_2, x_3)$ be a function of three variables, x_1, x_2, and x_3. The equation

$$F(x_1, x_2, x_3) = 0 \tag{73}$$

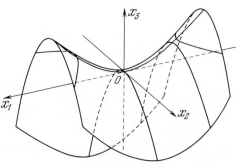

Fig. 39

defines a surface S in space, namely, the set of all points $x=(x_1,x_2,x_3)$ whose coordinates satisfy the equation (73).

Let Q be a certain plane in space and let L be an arbitrary curve in this plane. Draw lines $R(p)$ perpendicular to Q through every point p on the curve L. The set of all points in space that lie on one of the lines $R(p)$ is called *a cylinder*. The curve L is called *a directrix of the cylinder*, and each line $R(p)$ is called *a generator of the cylinder*. See Fig. 40. Suppose that Q is the coordinate plane $(1, 2)$ and that the curve L in this plane is given by the equation

$$F(x_1,x_2)=0. \tag{74}$$

Then the cylinder with directrix L in space also has the equation

$$F(x_1,x_2)=0. \tag{75}$$

Every equation of the form (75) is the equation of the cylinder with directrix L defined in the coordinate plane $(1, 2)$ by the equation (74).

Now let s be some fixed point in space and let S be a certain surface. Suppose that S has the property that for every point x on S different from s, all of the points on the line going through x and s also lie on the surface S. Then S is called *a cone with vertex s*. See the sketch in Fig. 41.

Let $F(x_1,x_2,x_3)$ be a homogeneous function of three variables having degree of homogeneity k. That is, F satisfies the condition

$$F(\alpha x_1,\alpha x_2,\alpha x_3)=\alpha^k F(x_1,x_2,x_3)$$

for all (x_1,x_2,x_3) in space and all real numbers α. It is clear that the equation

$$F(x_1,x_2,x_3)=0 \tag{76}$$

defines a cone with vertex 0.

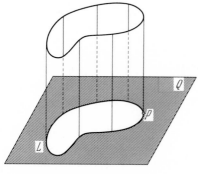

Fig. 40

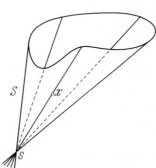

Fig. 41

3. Surfaces of Revolution. Let L be a curve in the $(1, 2)$ coordinate plane lying entirely above the first axis and defined by the equation

$$F(x_1, x_2) = 0.$$

Let us rotate the entire $(1, 2)$ plane about the first axis. The curve L will sweep out a surface S, which we call *a surface of revolution*. It is easy to see that an equation for the surface S is

$$F(x_1, +\sqrt{x_2^2 + x_3^2}) = 0.$$

The curve L can be chosen in any of the three coordinate planes, and we can rotate our coordinate plane about either coordinate axis in this coordinate plane, provided only that L lie entirely on one side of the axis in question. Corresponding equations for the surfaces of revolution can be written down at once.

We consider some interesting surfaces of revolution.

Let L be the straight line in the coordinate plane $(2, 3)$ and defined by the equation

$$k x_3 = x_2,$$

where k is a positive number. We can rotate this line about the third coordinate axis without ambiguity because its equation defines exactly one value of x_2 for each value of x_3. Doing this, we obtain the surface having as an equation

$$k x_3 = x_1^2 + x_2^2.$$

Squaring both sides, we get

$$k^2 x_3^2 = x_1^2 + x_2^2.$$

The surface S is a *circular cone* with vertex 0 and with axis the third coordinate axis.

Consider next the curve in the $(2, 3)$ plane with equation

$$x_3 = \alpha x_2^2,$$

where α is a nonzero constant. For definiteness, we will take this curve only for $x_2 \geq 0$. This curve is one half of a parabola in the $(2, 3)$ plane. Rotating it about the third coordinate axis, we obtain a surface S with equation

$$x_3 = \alpha(x_1^2 + x_2^2). \tag{77}$$

The surface S is called *a paraboloid of revolution*. It is very important to get a clear picture of its geometric form: see Fig. 42.

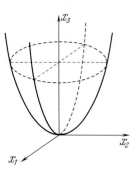

Fig. 42   Fig. 43

Next consider the curve L in the $(1, 2)$ plane with equation

$$\frac{x_1^2}{a^2} + \frac{x_2^2}{b^2} = 1.$$

Again, for the sake of definiteness we take only the part of the curve for which $x_2 \geq 0$. Plainly we have the upper half of an ellipse in the $(1, 2)$ plane. We do not require that $a \geq b$. If $a < b$, the foci of the ellipse lie on the second axis instead of the first. Rotating the curve L about the first axis, we obtain a surface of revolution with equation

$$\frac{x_1^2}{a^2} + \frac{x_2^2}{b^2} + \frac{x_3^2}{b^2} = 1. \tag{78}$$

This surface is called *an ellipsoid of revolution.* If $a > b$, the surface is stretched out along the axis of revolution. If $a < b$, then it is flattened along this axis. The geometric form of an ellipsoid of revolution is sketched in Fig. 43.

We now consider the surface obtained by rotating an hyperbola about an axis. Let us put our hyperbola in the $(1, 2)$ plane and describe it by the equation

$$\frac{x_1^2}{a^2} - \frac{x_2^2}{b^2} = 1.$$

We can rotate the hyperbola about the first coordinate axis (using only the upper half) to obtain a surface S_1. We can also rotate the hyperbola about the second coordinate axis (using only the right branch) to obtain a surface S_2. An equation for the surface S_1 is

$$\frac{x_1^2}{a^2} - \frac{x_2^2}{b^2} - \frac{x_3^2}{b^2} = 1. \tag{79}$$

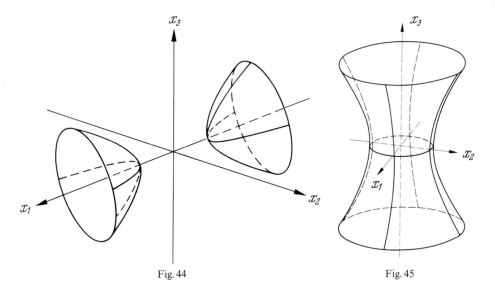

Fig. 44   Fig. 45

An equation for the surface S_2 is

$$\frac{x_1^2}{a^2}+\frac{x_2^2}{a^2}-\frac{x_3^2}{b^2}=1.\tag{80}$$

Again, it is very important to form a clear picture of both of the surfaces S_1 and S_2. See Figs. 44 and 45, respectively. The two surface differ radically from each other. The surface S_1 is called *an hyperboloid of revolution of two sheets*, since it consists of two separate parts. The surface S_2 is called *an hyperboloid of revolution of one sheet*, since it consists of only one piece. The equations for the surfaces S_1 and S_2 differ from each other in that the left side of (79) contains two negative signs, while the left side of (80) has but one.

4. Equations of Planes. Let Q be an arbitrary plane in space. Given a system of Cartesian coordinates in space, let us find an equation for Q. First draw a line R through the origin that is perpendicular to Q, and let q be the point of intersection of R and Q. Let a be the length of the line segment oq. Let e be a vector of length 1 lying on R. If $q\neq$ then e has the direction of oq. If $q=o$, then e may have either direction. We write e in coordinate form as

$$e=(\alpha_1,\alpha_2,\alpha_3).$$

It is clear that a point $x=(x_1,x_2,x_3)$ lies in the plane Q if and only if its projection onto the line R coincides with the point q. Citing formula (47) in the Supplement to Chapter I, we write this condition as

$$e\cdot x=a.$$

In coordinate form, this equality is written as

$$\alpha_1 x_1 + \alpha_2 x_2 + \alpha_3 x_3 = a. \tag{81}$$

We call this *the normal equation of a plane in space.* Note that

$$\alpha_1^2 + \alpha_2^2 + \alpha_3^2 = 1$$

and that the number a is nonnegative.

We revert for a moment to lines in a plane. The equation (81) has a complete analogue for an arbitrary line in a plane P. Every line has an equation

$$x \cos \varphi + y \sin \varphi = a, \tag{82}$$

where $e = (\cos \varphi, \sin \varphi)$ and a is a nonnegative number. The equation (82) is called *the normal equation of a line in the plane.*

5. Surfaces of the First and Second Orders. Let $F(x_1, x_2, x_3)$ be a polynomial in the three variables x_1, x_2, x_3 having either the first or the second degree. The equation

$$F(x_1, x_2, x_3) = 0$$

defines a surface in space. If the degree of F is one, this surface is said to be *of the first order,* and if the degree of F is two, the surface is said to be *of the second order.* We will prove in the Supplement to Chapter III that every surface of the first order is a plane. We will also in this Supplement present a complete classification of surfaces of the second order. A surface of the second order may be degenerate, that is, it may be a cylinder or a cone. These cases aside, every surface of the second order has one of the following forms (in a suitably chosen coordinate system).

1) *An Ellipsoid.* In a suitable coordinate system, the surface has an equation

$$\frac{x_1^2}{a_1^2} + \frac{x_2^2}{a_2^2} + \frac{x_3^2}{a_3^2} = 1. \tag{83}$$

Here a_1, a_2, and a_3 are positive numbers that determine the shape of the ellipsoid.

2) *An Hyperboloid of Two Sheets.* In a suitable coordinate system, the surface has an equation

$$\frac{x_1^2}{a_1^2} - \frac{x_2^2}{a_2^2} - \frac{x_3^2}{a_3^2} = 1. \tag{84}$$

Again a_1, a_2, and a_3 are positive numbers that determine the shape of the hyperboloid.

3) *An Hyperboloid of One Sheet.* In a suitable coordinate system, the surface has an equation

$$\frac{x_1^2}{a_1^2} - \frac{x_2^2}{a_2^2} + \frac{x_3^2}{a_3^2} = 1. \tag{85}$$

Again a_1, a_2, and a_3 are positive numbers that determine the shape of the hyperboloid.

4) *An Elliptic Paraboloid.* In a suitable system of coordinates, the surface has an equation

$$x_3 = \frac{x_1^2}{a_1^2} + \frac{x_2^2}{a_2^2}. \tag{86}$$

The numbers a_1 and a_2 are positive and define the shape of the paraboloid.

5) *An Hyperbolic Paraboloid.* In a suitable coordinate system, the surface has an equation

$$x_3 = \frac{x_1^2}{a_1^2} - \frac{x_2^2}{a_2^2}. \tag{87}$$

Once more a_1 and a_2 are positive numbers that determine the shape of the paraboloid.

To obtain a clear picture of the forms of the surfaces 1), 2), 3), and 4), it is useful to compare them with the surfaces of revolution defined by the equations (78)–(80) and (77). Each of the surfaces 1), 2), 3), 4) is obtained from one of our surfaces of revolution by a compression or stretching along a coordinate axis perpendicular to the axis of rotation.

The surface 5) can be compared with the surface S defined by equation (71), from which it is obtained by a compression or stretching either in the direction of the first coordinate axis or the second coordinate axis.

Chapter III
Analytic Geometry in the Plane

We pointed out in §5 that if we construct a system of Cartesian coordinates in a plane P and if $F(x, y)$ is a function of the two independent variables x and y, then the equation

$$F(x, y) = 0 \qquad (1)$$

defines a curve in the plane, namely, all of the points $z = (x, y)$ whose coordinates satisfy the equation (1).

Analytic geometry in the plane consists in the study of curves defined as in (1) for the cases in which $F(x, y)$ is a polynomial in the variables x and y of degree one or two. If $F(x, y)$ has degree one, then the corresponding curve is called a curve of the first order. If $F(x, y)$ has degree two, then the corresponding curve is called a curve of the second order. We will show that every curve of the first order is a straight line. We will also show that, degenerate cases aside, every curve of the second order is an ellipse, an hyperbola, or a parabola. The degenerate cases are those in which the equation (1) defines a pair of lines, a single line, a single point, or no points at all. We will prove all this in §10 of the present chapter. We have already presented, in §5, equations for ellipses, hyperbolas, and parabolas where the coordinate axes were chosen in a special way. To prove that the equation (1) defines an ellipse, an hyperbola, or a parabola, we may have to choose a a system of coordinates different from those in which we originally describe (1). We will do this in such wise that *in our new system of coordinates* the equation $F(x, y) = 0$ is changed into an equation of one of the forms presented in §5.

Our first task is therefore that of studying changes of coordinate systems in the plane. The following section, §9, is devoted to this task.

In §10, we show that all curves of the first order are straight lines and that all nondegenerate curves of the second order are ellipses, hyperbolas, or parabolas. In §11, the last of the main text, we consider the classical problem of finding all curves that can be obtained by intersecting a circular cone with a plane in space. We will show that all of these curves are in fact ellipses, hyperbolas, or parabolas (apart from degenerate cases). This fact was known to the ancient Greeks, at about the beginning of the Christian era. For this reason, ellipses, hyperbolas, and parabolas are called *conic sections*. We will give in §11 a beautiful geometric proof of this fact, that makes no use of the method of coordinates. This proof was discovered in the 19th century. Nowadays it seldom finds a place in mathematics courses. I include the proof in the present volume solely because of its geometric charm.

This chapter is devoted entirely to discussions of geometric facts. We will not use the results of this chapter in our subsequent study of analysis. The reader who is concerned mainly with analysis and not geometry may omit this chapter without harm. This applies particularly to the last section, §11. It will not be cited in the sequel, and furthermore, its results are not hard to obtain if we use the method of coordinates in three-dimensional space.

§9. Transformations of Cartesian Coordinates in the Plane

In §1, we constructed in explicit fashion a system of rectangular Cartesian coordinates in a plane P. The axis of abscissas was taken to be horizontal and was oriented from left to right. The axis of ordinates then had to be vertical, and we chose its orientation to be from below to above. We denoted the point of intersection of the two axes – the origin of coordinates – by the symbol o. We will call this system of coordinates in P the first system. We note as well that in the first system, we obtain the positive semi-axis of ordinates from the positive semi-axis of abscissas by rotating the latter through the angle $\frac{1}{2}\pi$ in the counterclockwise direction. We express this fact by saying that the first system of coordinates has *counterclockwise orientation* or is *positively oriented*. Let us now suppose that our plane is covered with transparent tracing paper and that we have copied our first system of coordinates on this tracing paper. Let us now move the tracing paper around in any way we like, but without lifting it from the plane. We will at once see two systems of coordinates: the first system is visible through the tracing paper and a second system that is drawn on the tracing paper. The situation is illustrated in Fig. 46.

Let o' denote the origin of the second system of coordinates. We obtain the second system of coordinates by moving the first system in the plane P. It is clear that the second system of coordinates is also oriented counterclockwise or positively.

We can also construct a second system of coordinates in the plane that is wholly independent of the first system. To do this, choose any straight line whatever in the plane, and give it an arbitrary direction. Take this line with its arbitrary direction for the axis of abscissas. As the axis of ordinates, take any line whatever that is perpendicular to the axis of abscissas, and give the axis of ordinates an arbitrary direction. Let us call this system of coordinates the second system: we use Fig. 46 once again to represent it. Let

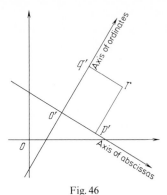

Fig. 46

o' be the origin of coordinates in the new system. We have chosen a direction on the axis of abscissas, and so the point o' divides the axis of abscissas into to semi-axes, the negative and positive semi-axes. Thus a point moving in the positive direction on the axis of abscissas goes from the negative semi-axis to the positive semi-axis. In just the same way, the origin o' divides the axis of ordinates into a negative semi-axis and a positive semi-axis.

Let r be an arbitrary point in our plane. Let us find its coordinates u and v in the second system of Cartesian coordinates. We drop a perpendicular rp' from r to the axis of abscissas in the second system of coordinates and also a perpendicular rq' from r to the axis of ordinates in the second coordinate system. (Again look at Fig. 46.) If p' lies on the positive semi-axis of abscissas, then the coordinate u is the length of the line segment $o'p'$. If p' lies on the negative semi-axis of abscissas, then the coordinate u is the negative of the length of the line segment $o'p'$. The ordinate v is defined analogously. To obtain the positive semi-axis of ordinates from the positive semi-axis of abscissas, we rotate the latter through an angle of $\frac{1}{2}\pi$ about the origin o'. This rotation can be either counterclockwise or clockwise. In the first case, we say that the second coordinate system is *oriented counterclockwise* or *positively*. In the second case, we say that the second system of coordinates is *oriented clockwise* or *negatively*. It is easy to exhibit a negatively oriented coordinate system in the plane. For example, take the first coordinate system and reverse the direction of the axis of ordinates. That is, take the direction on the axis of ordinates to be from above to below.

We recall that a system of coordinates obtained from the first system by moving it has positive orientation. It is easy to show that every positively oriented coordinate system is obtainable from the first system by moving it. In fact, choose any second system of coordinates that is positively oriented. We now move our tracing paper so that the origin drawn on it coincides with the new origin o'. Then rotate the tracing paper about the point o' until the axis of abscissas drawn on the tracing paper coincides with the axis of abscissas of our new coordinate system, and make sure also that their directions coincide. This is of course possible. The axis of ordinates drawn on the tracing paper will then coincide with the new axis of ordinates. The directions of these two axes coincides since both are positively oriented.

Below we will consider along with the first coordinate system a second system that is also positively oriented. That is, we suppose that we have two systems of coordinates in our plane P. The first is the original system as described in §1. The second is arbitrary, subject to the restriction that it have positive orientation. Let r be a point in the plane. We consider its coordinates in the first system of coordinates, writing

$$r = (x, y)_1.$$

We denote the coordinates of the same point r in the second coordinate system by (u, v), and we write specifically

$$r = (u, v)_2.$$

For given numbers u and v, the point r is determined and so also are its coordinates x and y. That is, the numbers x and y can somehow be expressed in terms of u and v: this under the assumption that we know the relative positions of the two coordinate systems.

Suppose that we have a curve K that is described by an equation (1) in the first system of coordinates. To find its equation in the second coordinate system, we must consider the function $F(x, y)$ and replace x and y by their expression in terms of u and v. Carrying out this replacement, we obtain a new function $G(u, v)$ and so an equation for the curve K in the second system of coordinates is

$$G(u, v) = 0.$$

Let us compute the coordinates x and y of a point r in the first coordinate system in terms of the coordinates u and v of this point in the second system of coordinates in two special cases.

Case 1. Suppose that the second coordinate system is obtained from the first by a parallel translation of the latter. Then the axis of abscissas in the second coordinate system is horizontal and is oriented from left to right. The axis of ordinates of the second coordinate system is vertical and is oriented from below to above (see Fig. 47). Consider the following three vectors in the plane:

$$z = o\,r, \quad w = o'\,r, \quad c = o\,o'.$$

It is clear that

$$z = c + w. \tag{2}$$

Let us find the coordinates of all three vectors z, w, and c in the first system of coordinates. We plainly have

$$z = (x, y)_1.$$

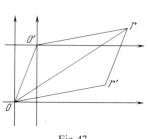

Fig. 47

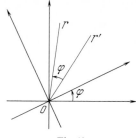

Fig. 48

The vector c describes the location of the second system of coordinates with respect to the first system. Let us write

$$c=(a, b)_1.$$

The coordinates of w in the second system are exactly

$$w=(u, v)_2.$$

To compute the coordinates of the vector w in the first coordinate system, we must construct the vector or', which is equal to the vector $o'r$, and determine the coordinates of the point r' in the first coordinate system. Again, see Fig. 47. Since the first and the second coordinate systems have parallel and similarly oriented axes, it is clear that the vector or' has just the same coordinates in the first system as the vector $o'r$ has in the second system. That is, we have

$$o'r=o'r=(u, v)_1.$$

Thus the vector equality (2) can be rewritten in the first coordinate system in the form of two equalities:

$$\left.\begin{array}{l} x=a+u, \\ y=b+v. \end{array}\right\} \tag{3}$$

The equalities (3) give the coordinates of a point r in the first coordinate system in terms of its coordinates in the second system.

Case 2. Suppose that the second coordinate system is obtained from the first by rotating the first system through an angle φ in the counterclockwise direction about the origin. Thus the points o and o' coincide. A sketch is given in Fig. 48. In our plane P, choose the point r' for which

$$r'=(u, v)_1.$$

(As before, we consider a point r for which $r=(u, v)_2$.) That is, r' is the point in the plane whose coordinates in the first system are equal to the coor-

dinates of r in the second system. From our definition of the second system, it is clear that the vector or is obtained from the vector or' by rotating or' through the angle φ in the counterclockwise direction. For convenience we now suppose that the plane P with the first system of coordinates is the plane of the complex variable z. We write

$$z = x + iy \quad \text{and} \quad w = u + iv. \tag{4}$$

That is, the complex number z represents the point r and the complex number w the point r'. Now define

$$\varepsilon = \cos \varphi + i \sin \varphi. \tag{5}$$

That is, ε is the complex number of modulus 1 and argument φ. Since the vector or is obtained from the vector or' by rotating or' through the angle φ (counterclockwise), it is clear that

$$z = w\varepsilon. \tag{6}$$

(Compare this with formula (30) of Chapter I.)

In the identity (6), substitute the expressions (4) and (5) for z, w, and ε. We find that

$$x + iy = (u + iv)(\cos \varphi + i \sin \varphi)$$
$$= (u \cos \varphi - v \sin \varphi) + i(u \sin \varphi + v \cos \varphi).$$

These equalities mean of course that

$$\left. \begin{array}{l} x = u \cos \varphi - v \sin \varphi, \\ y = u \sin \varphi + v \cos \varphi. \end{array} \right\} \tag{7}$$

The equalities (7) express the coordinates x and y of the point r in the first system of coordinates in terms of the coordinates u and v of the same point r in the second system.

The cases $\varphi = \frac{1}{2}\pi$ and $\varphi = \pi$ are of interest. For $\varphi = \frac{1}{2}\pi$, we obtain

$$\left. \begin{array}{l} x = -v, \\ y = u. \end{array} \right\} \tag{8}$$

For $\varphi = \pi$, we obtain

$$\left. \begin{array}{l} x = -u, \\ y = -v. \end{array} \right\} \tag{9}$$

In conclusion, we point out that the process of going from the first system of coordinates to any other positively oriented system can be carried out in two steps. First we carry out a parallel translation, as described in formula (3) and then a rotation as described in formula (7). We can also carry out these steps in the reverse order: first a rotation and then a parallel translation.

Example 1. In Example 1 of §4, Chapter I, we plotted the graph of the function

$$y = \frac{\gamma}{x},$$

γ being a positive constant. This is the same of course as plotting the curve defined by the equation

$$xy = \gamma. \tag{10}$$

Let us rotate our first system of coordinate axes through the angle $\frac{1}{4}\pi$. The equalities (7) become

$$x = 2^{-\frac{1}{2}}u - 2^{-\frac{1}{2}}v,$$

$$y = 2^{-\frac{1}{2}}u + 2^{-\frac{1}{2}}v.$$

Substituting these expressions for x and y in the equation (10), we obtain

$$\frac{u^2}{2\gamma} - \frac{v^2}{2\gamma} = 1.$$

This shows that the curve defined by the equation (10) is an hyperbola. Compare this with formula (26) of Chapter II.

Example 2. We pursue in more detail the remark made just before Example 1 about combining rotations and translations. Select an arbitrary positively oriented system of coordinates in the plane P. Let o' be its origin. We will designate this system as the third coordinate system and will write coordinates of a point r in this system as ξ and η (see Fig. 49):

$$r = (\xi, \eta)_3.$$

We now go from the first coordinate system to the third in two steps: first by a parallel translation and then by a rotation. The parallel translation carries the first origin o into the third origin o'. In the second coordinate system obtained by this parallel translation, we write the coordinates of a point r as u and v:

$$r = (u, v)_2.$$

The coordinates of a point r in the first system are denoted as usual by x and y:

$$r = (x, y)_1.$$

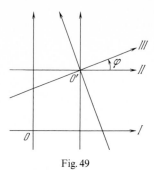

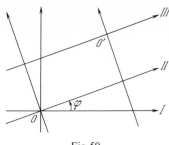

Fig. 49   Fig. 50

From formula (3) we get

$$\left.\begin{array}{l} x=a+u, \\ y=b+v. \end{array}\right\} \tag{11}$$

The second system of coordinates and the third have a common origin, namely ρ', and both have positive orientation. Thus the third is obtained from the second by rotation through an angle φ. Formula (7) shows that

$$u=\xi \cos \varphi - \eta \sin \varphi,$$
$$v=\xi \sin \varphi + \eta \cos \varphi.$$

Substituting these expressions for u and v in the equalities (11), we find the following equalities that relate the coordinates in the first and third systems:

$$\left.\begin{array}{l} x=a+\xi \cos \varphi - \eta \sin \varphi, \\ y=b+\xi \sin \varphi + \eta \cos \varphi. \end{array}\right\} \tag{12}$$

The numbers a, b, and φ describe the location and position of the third coordinate system relative to the first.

Let us now make the transition from the first coordinate system to the third by first carrying out a rotation and then a parallel translation. The process is illustrated in Fig. 50. The second system is obtained from the first by a rotation through the angle φ. Write the coordinates of a point r in the second system as u and v. Formula (7) gives

$$x=u \cos \varphi - v \sin \varphi,$$
$$y=u \sin \varphi + v \cos \varphi. \tag{13}$$

To obtain the third coordinate system from the second, we must translate the second system by the vector oo'. However, to use the formulas (3), we must write the coordinates of the vector oo' in the second coordinate system. Write these coordinates as

$$o\,o' = (\alpha, \beta)_2.$$

Formulas (3) then give

$$u = \alpha + \xi,$$
$$v = \beta + \eta.$$

Substitute these values for u and v in the equalities (13). This yields

$$x = \alpha \cos \varphi - \beta \sin \varphi + \xi \cos \varphi - \eta \sin \varphi,$$
$$y = \alpha \sin \varphi + \beta \cos \varphi + \xi \sin \varphi + \eta \cos \varphi. \tag{14}$$

The transformations (12) and (14) differ only in their form and are actually equivalent. This is because

$$a = \alpha \cos \varphi - \beta \sin \varphi,$$
$$b = \alpha \sin \varphi + \beta \cos \varphi.$$

We have thus established the connection between the third system of coordinates and the first system: we may use either (12) or (14). The first system of coordinates was described quite specifically in §1, Chapter I. This plays no essential rôle, however, since any of our sketches can always be rotated through an angle. Therefore formulas (12) and (14) establish the connection between any two positively oriented coordinate systems.

We finish our discussion of transformations of coordinates by considering an arbitrary *negatively* oriented system, which we will call the fourth system. A sketch is given in Fig. 51. Let ξ' and η' be the coordinates of a point r in the fourth system:

$$r = (\xi', \eta')_4.$$

We obtain the third system from the fourth by reversing the direction of the axis of ordinates in the fourth system. This reverses the orientation and so

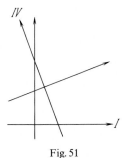

Fig. 51

gives a positively oriented system, that is, our third system. The coordinates (ξ, η) in the third system are given by the fomulas

$$\xi = \xi',$$
$$\eta = -\eta'. \tag{15}$$

Combining (15) and (12), we can write formulas for coordinates (x, y) of a point in the first system in terms of its coordinates in the fourth:

$$x = a + \xi' \cos \varphi + \eta' \sin \varphi,$$
$$y = b + \xi' \sin \varphi - \eta' \cos \varphi. \tag{16}$$

Thus the old coordinates (x, y) of a point r are given in terms of the new coordinates (ξ, η) or (ξ', η') by the formulas (12) if the new system is positively oriented and by the formulas (16) if the new system is negatively oriented. In both cases, the old coordinates x and y are polynomials of the first degree in the new coordinates.

In studying curves defined by equations of the form (1), we are interested only in properties of these curves that do not depend upon a special choice of coordinate systems. That is, we look for properties that do not change when x and y are transformed according to the formulas (12) or (16). Properties of a curve that alter under such transformations are not really geometric properties of the curve: they depend upon the position of the curve with respect to one or another coordinate system.

Suppose that $F(x, y)$ in (1) is a polynomial of degree n in the variables x and y. Then the new polynomials $G(\xi, \eta)$ and $G'(\xi', \eta')$ obtained from $F(x, y)$ by applying formulas (12) and (16) respectively are also polynomials in their variables whose degrees cannot exceed n. This is because x and y are polynomials in ξ and η (or ξ' and η') of the first degree. However, the degree of $G(\xi, \eta)$ (or $G'(\xi', \eta')$) cannot be smaller than n, since we can get $F(x, y)$ back from $G(\xi, \eta)$ (or $G'(\xi', \eta')$) by applying the inverse transformation of coordinates. If the degree of $G(\xi, \eta)$ (or $G'(\xi', \eta')$) were less than n, the degree of $F(x, y)$ would also be less than n. Therefore the degree of $F(x, y)$ is a geometric property of the curve, or, as we say, *an invariant*. The degree n of the polynomial $F(x, y)$ is called *the order of the curve defined by the equation* (1). As remarked above, analytic geometry in our sense is limited to the study of curves of the first and second orders.

§ 10. Curves of the First and Second Orders

In this section we will study curves defined by equations

$$F(x, y) = 0, \tag{17}$$

where $F(x, y)$ is a polynomial of the first or second degree. As was remarked in the introduction to the present chapter, the corresponding curve is called a *curve of the first (second) order* if $F(x, y)$ is a polynomial of the first (second) degree.

A curve of the first order is necessarily a straight line. We prove this first. A polynomial of the first degree in x and y has the form $a_1 x + a_2 y + a_0 = 0$, and so (17) in this case has the form

$$F(x, y) = a_1 x + a_2 y + a_0 = 0. \tag{18}$$

Suppose first that $a_2 \neq 0$. We divide (18) by a_0 and rewrite it in the form

$$y - y_0 = kx, \tag{19}$$

where

$$y_0 = -\frac{a_0}{a_2} \quad \text{and} \quad k = -\frac{a_1}{a_0}.$$

The equation (19) defines the straight line passing through the point $(0, y_0)$ and with slope k. (See formula (48) of Chapter II.) Suppose next that

$$a_2 = 0.$$

Since $F(x, y)$ has degree 1, a_1 in this case must be different from 0. Dividing (19) by a_1, we obtain the equation

$$x = x_0,$$

where

$$x_0 = -\frac{a_0}{a_1}.$$

Thus (18) defines a straight line in this case as well (see formula (49) of Chapter II).

Taking account once more of formulas (48) and (49) of Chapter II, we can say that straight lines are exactly those curves having equations of the first degree.

We now come to the considerably more complicated case of curves of the second order. We write the corresponding polynomial $F(x, y)$ in the form

$$F(x, y) = a_{11} x^2 + 2a_{12} xy + a_{22} y^2 + a_1 x + a_2 y + a_0.$$

We first rotate our coordinate system through an angle φ. That is, we write the coordinates (x, y) in terms of (u, v) according to the formulas (7). When this is done, the polynomial $F(x, y)$ becomes a polynomial $G(u, v)$.

We wish to choose the angle φ in such a way that the polynomial $G(u, v)$ does not contain the term uv. A simple computation shows that

$$G(u, v) = uv[-2a_{11} \cos \varphi \sin \varphi + 2a_{12}(\cos^2 \varphi - \sin^2 \varphi)$$
$$+ 2a_{22} \cos \varphi \sin \varphi] + \cdots.$$

In the last equality we have written only that term of the polynomial $G(u, v)$ that involves the term uv. We simplify the expression $[\cdots]$ in the last equation by using the well-known identities

$$\sin 2\varphi = 2 \sin \varphi \cos \varphi,$$
$$\cos 2\varphi = \cos^2 \varphi - \sin^2 \varphi.$$

The coefficient of uv in $G(u, v)$ is therefore equal to

$$(a_{22} - a_{11}) \sin 2\varphi + 2a_{12} \cos 2\varphi. \tag{20}$$

Let us find an angle φ for which the expression (20) is equal to zero. If

$$a_{12} = 0,$$

no rotation is needed: we take

$$\varphi = 0.$$

Suppose that $a_{12} \neq 0$. Setting (20) equal to zero, we find the following equation:

$$\operatorname{ctn} 2\varphi = \frac{a_{11} - a_{22}}{2a_{12}}.$$

From this equation, we can find an angle φ that makes (20) vanish. Having chosen such an angle, we write $G(u, v)$ as

$$G(u, v) = b_{11} u^2 + b_{22} v^2 + b_1 u + b_2 v + b_0. \tag{21}$$

Not both of the coefficients b_{11} and b_{22} can vanish. If they did, the polynomial $G(u, v)$ would be of degree 1 or would be a constant, and as pointed out at the end of §9 this is impossible since $F(x, y)$ has degree 2.

If we carry out a second rotation through the angle $\frac{1}{2}\pi$ (see formula (8)), the coefficients b_{11} and b_{22} will be interchanged. We will do this in the sequel wherever it is convenient.

We now subject our coordinate system to a parallel translation. The coordinates u and v are expressed in terms of the new coordinates ξ and η by the formulas

$$u = \alpha + \xi,$$
$$v = \beta + \eta \tag{22}$$

as in formula (3). Substitute the values from (22) into the expression (21) for $G(u, v)$. We obtain a polynomial $H(\xi, \eta)$:

$$H(\xi, \eta) = b_{11}\xi^2 + b_{22}\eta^2 + c_1\xi + c_2\eta + c_0,$$

where

$$c_1 = 2\alpha b_{11} + b_1,$$
$$c_2 = 2\beta b_{22} + b_2. \tag{23}$$

If $b_{11} \neq 0$, we can plainly choose the number α so that

$$c_1 = 0. \tag{24}$$

Namely, we set

$$\alpha = -\frac{b_1}{2b_{11}}.$$

Similarly, if $b_{22} \neq 0$, we can choose the number β so that

$$c_2 = 0. \tag{25}$$

We now take up the various cases given by different values of b_{11} and b_{22}.

Case 1. $b_{11} \neq 0$ and $b_2 \neq 0$. We choose α and β so that both (24) and (25) hold. Thus $H(\xi, \eta)$ has the form

$$H(\xi, \eta) = b_{11}\xi^2 + b_{22}\eta^2 + c_0. \tag{26}$$

If

$$c_0 \neq 0,$$

the equation of our curve can be rewritten in the form

$$-\frac{b_{11}}{c_0}\xi^2 - \frac{b_{22}}{c_0}\eta^2 = 1. \tag{27}$$

If the coefficients $-\dfrac{b_{11}}{c_0}$ and $-\dfrac{b_{22}}{c_0}$ have opposite signs, we may suppose with no loss of generality that the first of them is positive. (We have pointed out already that this can be done.) We then write

$$a = +\sqrt{-\frac{c_0}{b_{11}}}$$

and

$$b = +\sqrt{+\frac{c_0}{b_{22}}}.$$

The equation (27) becomes

$$\frac{\xi^2}{a^2} - \frac{\eta^2}{b^2} = 1. \tag{28}$$

In the new coordinates (ξ, η), the equation (28) is exactly equation (26) of Chapter II and so our curve is an hyperbola.

In the coefficients $-\dfrac{b_{11}}{c_0}$ and $-\dfrac{b_{22}}{c_0}$ are both positive, we will suppose that the first of them is less than or equal to the second. (As noted above, this involves no loss of generality.) We then define

$$a = +\sqrt{-\frac{c_0}{b_{11}}},$$

$$b = +\sqrt{-\frac{c_0}{b_{22}}}.$$

Our equation becomes

$$\frac{\xi^2}{a^2} + \frac{\eta^2}{b^2} = 1,$$

where $a \geq b$. In this case we have an ellipse (see formula (25) in Chapter II).

If both of the coefficients $-\dfrac{b_{11}}{c_0}$ and $-\dfrac{b_{22}}{c_0}$ are negative, there are no points (ξ, η) that satisfy equation (27). We thus encounter a degenerate case: there are no points at all on our curve.

Suppose now that in the right side of (26), we have

$$c_0 = 0.$$

Here too we are in a degenerate case, as we will prove. Equation (26) has the form

$$b_{11}\xi^2 + b_{22}\eta^2 = 0.$$

If both of the numbers b_{11} and b_{22} have the same sign, only the point $(0,0)$ will satisfy the equation: this is a degenerate case. If the numbers b_{11} and b_{22} are of different signs, we may suppose that b_{11} is positive and that b_{22} is negative. We may write

$$b_{11} = \frac{1}{a^2} \quad \text{and} \quad b_{22} = -\frac{1}{b^2}.$$

Equation (26) can then be written as

$$\frac{\xi^2}{a^2} - \frac{\eta^2}{b^2} = 0. \tag{29}$$

The left side of (29) can be factored, and so (29) can be rewritten as

$$\left(\frac{\xi}{a} - \frac{\eta}{b}\right)\left(\frac{\xi}{a} + \frac{\eta}{b}\right) = 0.$$

Each of the equations

$$\frac{\xi}{a} - \frac{\eta}{b} = 0,$$
$$\frac{\xi}{a} + \frac{\eta}{b} = 0 \tag{30}$$

defines a certain straight line in the plane P, and a point (ξ, η) lies on the curve defined by (29) if and only if it lies on one of these lines. Thus the curve is actually a pair of intersecting lines and again is a degenerate case.

Case 2. Exactly one of the number b_{11} and b_{22} is different from 0. We may suppose, again with no loss of generality, that $b_{22} \neq 0$. We then choose the number β so that

$$c_2 = 0$$

(see (23)) and we take the number α to be 0. Thus we have

$$\xi = u,$$

and after our second transformation of coordinates we obtain

$$H(u, \eta) = b_{22}\eta^2 + c_1 u + c_0.$$

Suppose that

$$c_1 = 0.$$

Let us show that we have a degenerate case. The equation of our curve is

$$\eta^2 = -\frac{c_0}{b_{22}}. \tag{31}$$

If the right side is positive, we find that

$$\eta = \pm\sqrt{-\frac{c_0}{b_{22}}}.$$

Thus the equation (31) defines two lines parallel to the axis of abscissas. If the right side of (31) is equal to 0, then we get

$$\eta^2 = 0,$$

and our equation defines a single line, namely the axis of abscissas. If the right side of (31) is negative, there are no points whose coordinates satisfy equation (31).

Suppose finally that

$$c_1 \neq 0.$$

We set

$$u = \alpha + \xi.$$

The equation of our curve becomes

$$b_{22}\eta^2 + c_1\xi + c_1\alpha + c_0 = 0.$$

We can choose α so that

$$c_1\alpha + c_0 = 0,$$

so that the equation of our curve has the form

$$b_{22}^2 + c_1 = 0. \tag{32}$$

If the numbers b_{22} and c_1 have the same sign, we rotate our system of coordinates through the angle π. Then both ξ and η change their signs. Again, retaining our notation, we suppose that b_{22} and c_1 have different signs. The equation (32) can be rewritten in the form

$$\eta^2 = -\frac{c_1}{b_{22}}\xi,$$

or

$$\eta^2 = 2f\xi,$$

where f is positive. Therefore our curve is a parabola (see formulas (27) of Chapter II).

Therefore a curve of the second order, degenerate cases apart, is an ellipse, an hyperbola, or a parabola.

Example 3. We give closer attention to the hyperbola with equation

$$\frac{x^2}{a^2} - \frac{y^2}{b^2} = 1. \tag{33}$$

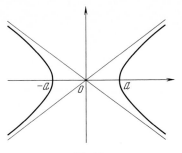

Fig. 52

Since this curve is symmetric with respect to both the axis of abscissas and the axis or ordinates, if suffices to study the behavior of the curve in the first quadrant. For this purpose, we solve for y in terms of x, with the agreement that both x and y be positive. We find

$$y = +b\sqrt{\frac{x^2}{a^2} - 1}.$$

For $x < a$, the right side of the last equality is a pure imaginary, and so there are no points with positive abscissa less than a that lie on the hyperbola (see Fig. 52). For $x = a$, there is only one point, namely $(a, 0)$ with abscissa a, that lies on the hyperbola. This obviously is the intersection of the hyperbola with the axis of abscissas. Now as x increases without bound, the value of y also increases without bound. To get some idea of how y increases with x, let us compare y with the behavior of points on the line

$$\hat{y} = \frac{b}{a}x, \tag{34}$$

which goes through the origin and has slope $k = \frac{b}{a}$. We denote the ordinate on this line by the symbol $\hat{y}$ to distinguish it from the ordinate of a point on the hyperbola.

Consider the difference

$$\hat{y} - y = \frac{b}{a}(x - \sqrt{x^2 - a^2}).$$

Multiply and divide the right side of this expression by the function $x + \sqrt{x^2 - a^2}$. We find that

$$\hat{y} - y = \frac{ab}{x + \sqrt{x^2 - a^2}}.$$

From the last equality, it is clear that $\hat{y}-y$ goes to zero as x gets larger and larger. The straight line (34) is called *an asymptote of the hyperbola* (33). The behavior we find in the first quadrant is repeated in the other three. Thus our hyperbola has two straight lines as asymptotes. They can be described by the single equation

$$\frac{x^2}{a^2}-\frac{y^2}{b^2}=0.$$

Both branches of the hyperbola get arbitrarily close to both of these lines, in the appropriate quadrants (see Fig. 52).

Example 4. Let us study the function

$$F(x, y)=a_{11}x^2+2a_{12}xy+a_{22}y^2. \tag{35}$$

This function of the two variables x and y is called *a quadratic form in the variables x and y.* Quadratic forms in many variables play a great rôle in algebra and its applications. Here we will obtain some properties of quadratic forms in two variables.

We can consider a quadratic form $F(x, y)$ as a function of the vector $z=(x, y)$, that is

$$F(z)=F(x, y). \tag{36}$$

That is, we associate with each point z in the plane a specified number $F(z)$. The number $F(z)$ must be independent of the coordinate system that we use: it depends only on the point. To compute the values of $F(z)$ we must choose some fixed coordinate system and compute the function $F(z)$ in this coordinate system by the formula (36). We can choose a coordinate system, ordinarily different from the original system in which the quadratic form is defined as in (35), in which we get

$$F(z)=G(u, v)=b_{11}u^2+b_{22}v^2 \tag{37}$$

(see formula (21)). It is easier to study the quadratic form $F(z)$ in this system of coordinates than in the original system, since the term in uv is absent.

Let us study the quadratic form $F(z)$ for vectors ε of length 1. We pose the problems of finding the vector ε_1 of length 1 where the function $F(\varepsilon)$ assumes its greatest value, or as one says, its *maximum*, and of finding the vector ε_2 where the function $F(\varepsilon)$ assumes its least value or *minimum*. It is convenient to solve this problem in the second coordinate system, where the quadratic form $F(z)$ has the form (37). Every vector of length 1 in the second coordinate system can be written in the form

$$\varepsilon=(\cos\alpha, \sin\alpha),$$

where α is the angle of inclination of the vector ε with the positive semi-axis of abscissas in the second system of coordinates. Putting these values for (u, v) in formula (37), we obtain

$$F(\varepsilon) = b_{11} \cos^2 \alpha + b_{22} \sin^2 \alpha. \tag{38}$$

If the numbers b_{11} and b_{22} are equal to each other, say both are equal to b, we find that

$$F(\varepsilon) = b$$

for all vectors ε of length 1. That is, the function $F(\varepsilon)$ assumes both its maximum and minimum values at every vector ε.

Suppose next that b_{11} and b_{22} are different. We will suppose that

$$b_{11} > b_{22}.$$

We rewrite the formula (38) as

$$F(\varepsilon) = b_{11} - (b_{11} - b_{22}) \sin^2 \alpha.$$

We see that $F(\varepsilon)$ assumes its maximum at exactly two vectors: the vector ε_1 with $\alpha = 0$ and the vector ε_1' with $\alpha = \pi$. Note that $\varepsilon_1' = -\varepsilon_1$. The function $F(\varepsilon)$ assumes its minimum value also at two vectors: the vector ε_2 with $\alpha = \frac{1}{2}\pi$ and its negative, the vector ε_2' with $\alpha = \frac{3}{2}\pi$. Finally note that the vectors ε_1 and ε_2 are perpendicular to each other.

§11. Conic Sections

In this section, we carry out the program promised in the introduction to the present chapter. Namely, we will show that ellipses, hyperbolas, and parabolas can be obtained as intersections of a circular cone (conical surface) with a plane not going through the vertex of the cone.

We have already defined general cones in the Supplements to Chapter II. For the sake of definiteness, we give the definition of *a circular conical surface* here in explicit form (see Fig. 53). Consider a horizontal plane in three-dimensional Euclidean space and a circle N in this plane with center at a point c. From c we draw a line perpendicular to the plane and directed upward from the plane. Choose a point s on this line. For every point p on the circle N, we draw the straight line containing p and s, and extending infinitely in both directions. *The circular cone C* is the surface consisting of all points that lie on one of these lines. Each of the lines is called *a generator of the cone C* and the point s is called its *vertex*. Every generator of C is divided by the vertex into two *semigenerators*, one lying above c and the other

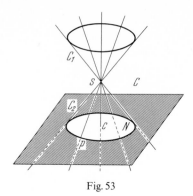

Fig. 53

below. The points of the cone that lie on upper semigenerators comprise *the upper nappe of the cone;* we designate this nappe as C_1. *The lower nappe C_2 is defined similarly.* The vertical line (extending infinitely in both directions) that passes through c and s is called *the axis of the cone.* Let γ denote the angle between the generators of the cone and its axis. The angle γ determines the shape of the cone.

Consider a spherical surface S with its center on the axis of the cone. We say that S is *inscribed in the cone C* if it is tangent to the surface of the cone on some circle L. Such a sphere S must lie completely within one of the nappes of the cone.

We now consider a plane P in our three-dimensional space that does not pass through the vertex s of our cone. The plane P intersects the cone in a certain curve K. Our aim is to identify all possible forms that the curve K can assume. To do this, we first construct a plane P' parallel to the plane P and passing through the vertex s. Three distinct cases can arise.

Case 1. The plane P' intersects C only at its vertex. Then the nappes C_1 and C_2 lie on opposite sides of P', and the plane P intersects only one of the nappes. For definiteness we will suppose that P intersects the nappe C_1. We will prove below that the curve K in this case is an ellipse.

Case 2. The plane P' intersects the cone C in two distinct generators of C. In this case, the plane P intersects both of the nappes C_1 and C_2. We will prove below that the curve K in this case is an hyperbola (both branches).

Case 3. The plane P' intersects the cone C in exactly one generator. Call this generator M. Thus P' is tangent to the cone C along the generator M, and therefore the plane P, which does not contain M, intersects only one nappe of the cone. Again for definiteness we suppose that P intersects only the nappe C_1. We will prove below that the curve K in this case is a parabola.

We proceed to the proofs.

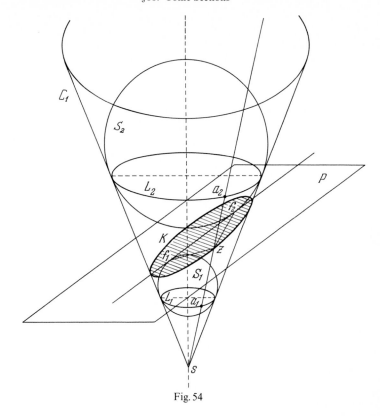

Fig. 54

Case 1. Here the plane P intersects every generator of C, and in view of our hypothesis, it intersects the upper half of each generator (see Fig. 54). To prove that K is an ellipse, we inscribe two spheres S_1 and S_2 in the upper nappe C_1. The first sphere, S_1, lies below the plane P and is tangent to the plane P from below. The sphere S_1 is indicated in Fig. 54. Let L_1 be the circle in which S_1 intersects the nappe C_1, and let f_1 be the point of tangency of S_1 with the plane P. The sphere S_2 is constructed similarly, except that it lies above the plane P. It is inscribed in the nappe C_1 and is tangent to the plane P from above P. The sphere S_2 is also indicated in Fig. 54. Let L_2 be the circle in which S_2 intersects C_1 and let f_2 be the point at which S_2 is tangent to P. Now consider an arbitrary point z on the curve K: note that the entire curve K lies between the circles L_1 and L_2 on the nappe C_1. Through the point z, there passes exactly one generator of C. Let a_1 be the point where this generator intersects the circle L_1 and let a_2 be the point where this generator intersects the circle L_2. For clarification, see Fig. 54. Plainly the length of the line segment $a_1 a_2$ does not depend on the choice of the point z on K, since L_1 and L_2 lie in planes parallel to the original plane used in defining C. Let $2a$ denote the length of the line segment $a_1 a_2$. It is obvious that

$$l(z, a_1) + l(z, a_2) = 2a. \tag{39}$$

We now draw line segments from z to the points f_1 and f_2. The line segments zf_1 and za_1 are both tangent to the sphere S_1 from the point z. Therefore they have equal lengths:

$$l(z, f_1) = l(z, a_1). \tag{40}$$

Exactly the same reasoning applies to the line segments za_2 and zf_2:

$$l(z, f_2) = l(z, a_2). \tag{41}$$

Combining the equalities (39), (40), and (41), we find that

$$l(z, f_1) + l(z, f_2) = 2a.$$

As already noted, the number a does not depend upon the choice of the point z on K. Therefore the curve K is the ellipse with foci at f_1 and f_2 for which the sum of the distances from f_1 and f_2 is $2a$. (The reader may refer to formula (14) of Chapter II.) Thus Case 1 is complete.

Case 2. We sketch the position of the plane P in Fig. 55. The plane P intersects both nappes of C. Let K_1 be the curve in which P intersects C_1 and K_2 the curve in which P intersects C_2. Thus K consists of two separated pieces. We will prove that K is an hyperbola with branches K_1 and K_2. As in our study of Case 1, we use inscribed spheres. The sphere S_1 is inscribed in the nappe C_1 and is tangent to the plane P. There is only one choice for the sphere S_1. Let L_1 be the circle in which S_1 intersects C_1 and let f_1 be the point at which S_1 is tangent to P. See Fig. 55 once more. The sphere S_2 is constructed similarly. It is inscribed in the lower nappe C_2, is tangent to the plane P at the point f_2, and intersects the nappe C_2 in a circle L_2. A sketch appears in Fig. 55. It is easy to see that the curve K_1 lies above the circle L_1 and that the curve K_2 lies below the circle L_2. Let z be an arbitrary point on the curve K_1. Consider the generator of C passing through z. Let a_1 be the point where this generator intersects L_1 and a_2 the point where it intersects L_2. The length of the line segment $a_1 a_2$ again is independent of the choice of z. Let $2a$ be the length of the line segment $a_1 a_2$. Since z lies above the circle L_1, it is clear that

$$l(z, a_2) - l(z, a_1) = 2a. \tag{42}$$

As in Case 1, we draw line segments from z to the points f_1 and f_2. Again as in Case 1, we see that zf_1 and za_1 are of equal length, as are zf_2 and za_2:

$$l(z, f_1) = l(z, a_1) \tag{43}$$

and

$$l(z, f_2) = l(z, a_2). \tag{44}$$

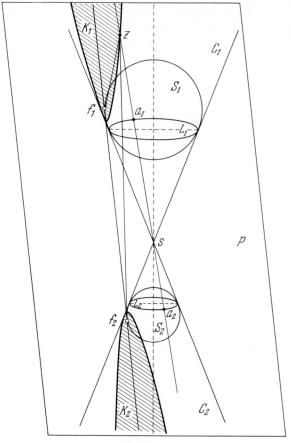

Fig. 55

Combining the equalities (42)–(44), we find that

$$l(z, f_1) - l(z, f_2) = 2a.$$

Citing formula (15) of Chapter II, we see that the curve K_1 is one branch of the hyperbola with foci f_1 and f_2 and with difference of distances from the foci equal to $2a$. The argument for the branch K_2 is very similar, and we omit it. Thus Case 2 is complete.

Case 3. The plane P is parallel to a plane P' that is tangent to the cone C. Let M' be the generator of C in which P' intersects C. In this case, we use only one inscribed sphere, denoted by S_0. The sphere S_0 is inscribed in the upper nappe C_1 and is tangent to the plane P. Let L_0 be the circle in which S_0 intersects C_1 and let f_0 be the point of tangency of S_0 and the plane P. Let Q be the plane containing the circle L_0 and let D' denote the

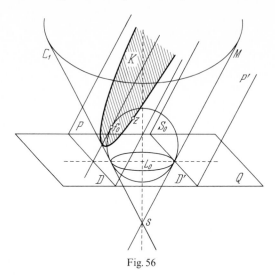

Fig. 56

line of intersection of the planes Q and P'. It is easy to see that the lines D' and M' are perpendicular to each other. Let D be the line of intersection of the planes Q and P. All of this construction is shown in Fig. 56. Obviously the lines D and D' are parallel. Now let z be an arbitrary point of the curve K. Let R be the plane containing z and the generator M' of C. We make a sketch in Fig. 57 of the plane R, in which we carry out further constructions. Let E be the line in which the planes R and Q intersect. Since the point z lies on the cone C, the plane R intersects C not only along the generator M' but also along a second generator M'', namely, the generator on which z lies. Observe now that the generators M' and M'' form equal angles with the line E in the plane Q. Since the point z lies in the plane P, the line of intersection M of the planes R and P contains z and is parallel to M'. Therefore the lines M and M'' form equal angles with the line E. Let a and a'' be the points of intersection of the lines M and M'' respectively with the line E. We thus obtain an isosceles triangle $z\,a\,a''$ with equal angles at the point a and a''. Therefore the lengths of the line segments $z\,a$ and $z\,a''$ are equal.

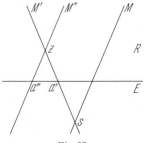

Fig. 57

We have already observed that the lines D' and M', lying in the plane P', are perpendicular. Therefore the lines D and M, which are parallel to D' and M' and lie in the plane P, are also perpendicular. The line segments $z f_0$ and $z a''$ are tangent to the sphere S_0 at the points f_0 and a'' respectively. Hence they have the same length. Since the length of the line segment $z a$ is equal to the length of the line segment $z a''$, we have proved that

$$l(z, f_0) = l(z, a).$$

This equality holds for all points z on the curve K. Therefore K is a parabola with focus f_0 and directrix D: see the definition of a parabola in §5. This completes our analysis of Case 3.

Example 5. We have shown that every conic section (apart from degenerate cases) is an ellipse, an hyperbola, or a parabola. Ellipses can have an infinite number of shapes: an ellipse is defined by the two parameters a and b, as in formula (25) of Chapter II. Similarly, an hyperbola is defined by two parameters a and b, as in formula (26) of Chapter II. A parabola, on the other hand, is defined by a single parameter f, as in formula (27) of Chapter II.

One may ask whether or not *every* ellipse, hyperbola, and parabola can be obtained as a conic section. To answer this question, we construct along with the plane P a plane P_1 parallel to P and also not passing through the vertex s of our cone C. Let K_1 be the curve in which P_1 intersects C. It is clear that the curves K and K_1 can be obtained from each other by a dilation, as in Example 2 of §5, Chapter II. To determine the coefficient α of this dilation, draw a line through the vertex s that is perpendicular to the planes P and P_1, and intersecting these planes in the points p and p_1 respectively. It is evident that the coefficient of dilation α is given by the formula

$$\alpha = \frac{l(s, p_1)}{l(s, p)}.$$

Thus we obtain along with a curve K all of the curves that can be obtained from K by dilations.

It follows immediately that all parabolas are conic sections, since every parabola can be obtained from a given parabola by a dilation (again see Example 2 of §5 in Chapter II).

The corresponding assertion for ellipses and hyperbolas is not obvious at first blush, however, since the ratio $\dfrac{b}{a}$ is preserved under dilations for both ellipses and hyperbolas. We may try to make this ratio whatever we wish by rotating the plane P.

The attempt is successful for ellipses. For a plane P perpendicular to the axis of the cone, the curve K is a circle, that is, we get $\dfrac{b}{a} = 1$. As we rotate the plane P so that its angle of inclination with the axis decreases, and also move the point of intersection of P with the axis farther and farther from the vertex, we see that the ratio $\dfrac{b}{a}$ can be made as small as we like. That is, every ellipse can be obtained as the intersection of a given cone with a certain plane P.

The situation with hyperbolas is more complicated. Suppose that we have a given cone C with a specified angle γ as defined at the beginning of the present section. Let P be a plane whose intersection K with C is an hyperbola, and let P' be the plane parallel to P that goes through the vertex s of C. The plane P' intersects C in two distinct generators. It is easy to see that these generators are the asymptotes of K. The largest possible angle between the asymptotes is obtained when P' contains the axis of the cone. In this case the angle is 2γ. For the hyperbola K, we see that the largest possible value of $\dfrac{b}{a}$ is $\tan\gamma$. For our given cone, all values of $\dfrac{b}{a}$ less than $\tan\gamma$ are also attainable. Thus, to get all possible hyperbolas, we have to take cones with larger and larger angles γ.

We have thus demonstrated that every ellipse, hyperbola, and parabola is a conic section.

Supplement to Chapter III

Here we take up the following: 1) transformation of coordinates in (three-dimensional) space; 2) classification of surfaces of the first and second orders; 3) conic sections treated by coordinates in space.

1. Transformation of Coordinates. Consider the rectangular Cartesian system of coordinates in space that we defined in the Supplement to Chapter I. We will take this as our first system and will construct an arbitrary second system. Let o' be the origin of the new system. For axes in the new system we take any three mutually perpendicular lines going through o', and provided with arbitrary directions. We define coordinates of a point r in the second system just as we did for the first system. That is, we project the point r on the three new axes, obtaining points p'_1, p'_2, and p'_3. The coordinates u_1, u_2, u_3 of r in the new coordinate system are the lengths of the intervals $o'p'_1$, $o'p'_2$, and $o'p'_3$ each with the appropriate sign. We write r in the second coordinate system as

$$r = (u_1, u_2, u_3)_2.$$

Let x_1, x_2, x_3 be the coordinates of r in the first system. We write

$$r = (x_1, x_2, x_3)_1.$$

We pose the problem of writing the coordinates x_1, x_2, x_3 in terms of u_1, u_2, u_3. Let us solve this problem in the two simplest cases.

Case 1. The axes of the second system are parallel to the axes of the first system and have the same direction. Let c be the vector oo'. We write

$$c = (c_1, c_2, c_3)_1,$$

meaning by this of course that c has coordinates c_1, c_2, c_3 in the first system. Just as was done in the plane in §9, we find that

$$x_1 = c_1 + u_1,$$
$$x_2 = c_2 + u_2,$$
$$x_3 = c_3 + u_3.$$

We write these equalities in shortened form as

$$x_i = c_i + u_i \quad (i = 1, 2, 3).$$

Thus we write the coordinates of a point r in the first system in terms of the coordinates of r in the second system.

Case 2. The origin o' of the second system is the same as the origin o of the first system. Along with our point r in space, we consider another point s for which

$$s = (y_1, y_2, y_3)_1 \quad \text{and} \quad s = (v_1, v_2, v_3)_2.$$

The scalar product rs of the vectors r and s (which was defined on page 27 of the Supplement to Chapter I) has a geometric interpretation that is independent of the particular choice of coordinate systems. This is:

$$r \cdot s = |r| \cdot |s| \cos \gamma,$$

where γ is the angle between the vectors r and s. On the other hand, the scalar product can be written in both of our coordinate systems:

$$r \cdot s = x_1 y_1 + x_2 y_2 + x_3 y_3 = u_1 v_1 + u_2 v_2 + u_3 v_3.$$

This formula is of fundamental importance. It states that the scalar product of two vectors is expressed by exactly the same formula in any two coordinate systems having the same origin.

Let e_1, e_2, e_3 be vectors of unit length that lie in the positive directions on the three coordinate axes of the first system of coordinates. We now write

$$e_i = (\alpha_{i,1}, \alpha_{i,2}, \alpha_{i,3})_2 \quad (i = 1, 2, 3). \tag{45}$$

That is, we write the vectors e_i in the second system of coordinates. For our point r, we plainly have

$$x_i = r \cdot e_i \quad (i = 1, 2, 3).$$

(Compare this with formula (46) in the Supplement to Chapter I.) Now write the scalar products on the right side of the last equalities in the second system of coordinates. We obtain

$$x_i = \alpha_{i,1} u_1 + \alpha_{i,2} u_2 + \alpha_{i,3} u_3 \quad (i = 1, 2, 3). \tag{46}$$

Thus (46) expresses the coordinates of the point r in the first system by means of its coordinates in the second system.

In order to compute the coordinates of a point in the first coordinate system from its coordinates in the second system, it thus suffices to know the nine numbers $\alpha_{i,j}$ $(i, j = 1, 2, 3)$. In the branch of mathematics known as linear algebra, it is customary to write these coefficients in the form of a square array:

$$\begin{pmatrix} \alpha_{1,1} & \alpha_{1,2} & \alpha_{1,3} \\ \alpha_{2,1} & \alpha_{2,2} & \alpha_{2,3} \\ \alpha_{3,1} & \alpha_{3,2} & \alpha_{3,3} \end{pmatrix}. \tag{47}$$

Such a square array of numbers is called *a matrix*. The first row of the matrix (47) consists of the coefficients for writing x_1 in terms of u_1, u_2, u_3; and similarly for the second and third rows. If we adopt the same procedure in the case of transformation of coordinates in the plane, we obtain the following matrix:

$$\begin{pmatrix} \cos \varphi & -\sin \varphi \\ \sin \varphi & \cos \varphi \end{pmatrix}. \tag{48}$$

Plainly the elements of the matrix (48) are not independent of each other but all depend upon only one parameter φ, the angle of rotation. In much the same way, the nine entries in the matrix (47) are not independent of each other. They can be expressed as functions of three angles, but in a rather awkward way. Therefore the relations between the elements of the matrix (47) are usually expressed in another way, by means of equations that connect these entries. These relations are easy to obtain if we use certain properties of the vectors e_1, e_2, and e_3, namely, each of them has length 1 and every pair of them are perpendicular.

In terms of scalar products, we write

$$\left. \begin{aligned} e_1^2 = e_2^2 = e_3^2 = 1, \\ e_1 \cdot e_2 = e_1 \cdot e_3 = e_2 \cdot e_3 = 0. \end{aligned} \right\} \tag{49}$$

We customarily write all of these equalities in abbreviated form:

$$e_j \cdot e_k = \delta_{j,k}, \tag{50}$$

where $\delta_{j,k}$ is *Kronecker's δ-symbol*, namely the number that is 1 if $j=k$ and is 0 if $j \neq k$. To obtain the relations among the entries of the matrix (47), we need only to rewrite (50) using the expressions (45) for the coordinates of the vectors e_i in the second coordinate system. We leave the details to the reader.

Suppose now that we have a third coordinate system in space with origin o' and with arbitrarily oriented axes of coordinates going through o'. We write the coordinates of a point r in this third system of coordinates as

$$r = (\xi_1, \xi_2, \xi_3)_3.$$

Regard this transformation of coordinates as being made first by a parallel translation as in Case 1 and then by a transformation as in Case 2. We find that

$$x_i = c_i + \alpha_{i,1}\xi_1 + \alpha_{i,2}\xi_2 + \alpha_{i,3}\xi_3 \quad (i=1,2,3). \tag{51}$$

Note that the formulas (51) express the original coordinates of a point in terms of the coordinates in the third system as polynomials of the first degree in these coordinates. The first and the third coordinate systems have just the same status, and so we can also express the coordinates ξ_1, ξ_2, ξ_3 in terms of x_1, x_2, x_3 also as polynomials of the first degree.

Suppose now that we are given a certain surface in space, described in the first system of coordinates by

$$F(x_1, x_2, x_3) = 0, \tag{52}$$

where $F(x_1, x_2, x_3)$ is a polynomial of degree n. In the third system of coordinates, we can describe this surface by replacing x_i in (52) by its values in terms of the coordinates ξ_j given by (51). This gives us a polynomial $G(\xi_1, \xi_2, \xi_3)$. In the third coordinate system the surface is given by the equation

$$G(\xi_1, \xi_2, \xi_3) = 0.$$

As in §9, we see that the polynomial $G(\xi_1, \xi_2, \xi_3)$ also has degree n. Thus the degree n is a geometric invariant of the surface. We call it *the order of the surface*.

2. Classification of Surfaces of Orders 1 and 2. Again we suppose that we have a coordinate system in space as described in the Supplement to Chapter I. Let $F(x_1, x_2, x_3)$ be a polynomial of degree 1 or 2. We will classify the surfaces

$$F(x_1, x_2, x_3) = 0. \tag{53}$$

For a polynomial $F(x_1, x_2, x_3)$ of degree 1, so that the surface (53) is of order 1, we will see that the surface is a plane. If $F(x_1, x_2, x_3)$ has degree 2, so that the surface (53) has order 2, we will show that the surface is degenerate (that is, is a cylinder or a cone) or has one of the five forms listed in the Supplement to Chapter II (see formulas (83)–(87)).

Case a). A polynomial F of degree 1 has the form

$$F(x_1, x_2, x_3) = a_1 x_1 + a_2 x_2 + a_3 x_3 + a_0.$$

At least one of the numbers a_1, a_2, a_3 must be different from 0. There exists a number β such that the polynomial

$$\beta F(x_1, x_2, x_3) = b_1 x_1 + b_2 x_2 + b_3 x_3 + b_0$$

has the following properties. First, the number b_0, which we write as $-b$, is nonpositive. Second, the numbers b_1, b_2, b_3 have the property that

$$b_1^2 + b_2^2 + b_3^2 = 1.$$

In this case, the equation (53) can be rewritten as

$$b_1 x_1 + b_2 x_2 + b_3 x_3 = b.$$

This equation is the normal equation of a plane (see formula (81) of the Supplement to Chapter II). This proves that a surface of order 1 is a plane. We note also that every plane is a surface of order 1.

Case b). Consider first a polynomial $F(x_1, x_2, x_3)$ of degree 2 containing no terms of the first degree and containing no constant term. Such a polynomial is called *a quadratic form in the three variables* x_1, x_2, x_3. Let $r = \varepsilon$ be an arbitrary vector of length 1. Let ε_1 be a vector of length 1 at which the function $F(\varepsilon)$ assumes its maximum. Now choose a new system of coordinates. The new origin is the old origin o. The first axis contains ε_1 and has the direction ε_1. The second and third axes may be arbitrary: they need only be perpendicular to each other and to the first axis. In this second system of coordinates, we write

$$r = (u_1, u_2, u_3)_2.$$

Using (46), we write the quadratic form $F(x_1, x_2, x_3)$ in terms of u_1, u_2, u_3, obtaining

$$F(r) = G(u_1, u_2, u_3).$$

Suppose now that the vector r belongs to the (1, 2) coordinate plane for the second system of coordinates. We then have

$$F(r) = b_{11} u_1^2 + 2 b_{12} u_1 u_2 + b_{22} u_2^2. \tag{54}$$

For all vectors ε of length 1, the quadratic form $F(\varepsilon)$ attains its maximum for $\varepsilon = \varepsilon_1$. In particular, for vectors of length 1 in the $(1, 2)$ coordinate plane in the second coordinate system, $F(\varepsilon)$ assumes its maximum at the vector ε_1. From Example 4 in Chapter III, we infer that the quadratic form (54) lacks a mixed term. That is, (54) has the form

$$F(r) = b_{11} u_1^2 + b_{22} u_2^2.$$

We next consider $F(r)$ for vectors in the $(1, 3)$ coordinate plane in the second system of coordinates. The argument just given shows that $F(r)$ lacks a term in $u_1 u_3$. Therefore the quadratic form $F(r)$ has the form

$$F(r) = b_{11} u_1^2 + b_{22} u_2^2 + 2 b_{23} u_2 u_3 + b_{33} u_2^3.$$

For vectors r belonging to the $(2, 3)$ coordinate plane in the second coordinate system, $F(r)$ has the form

$$F(r) = b_{22} u_2^2 + 2 b_{23} u_2 u_3 + b_{33} u_3^2. \tag{55}$$

We have already shown that for quadratic forms of the form (55) in the two variables u_2 and u_3, one can choose a system of coordinates in the $(2, 3)$ coordinate plane so that in the new variables ξ_2, ξ_3 the form $F(r)$ has the form

$$F(r) = c_{22} \xi_2^2 + c_{33} \xi_3^2.$$

(See formula (37).) Let us keep the first axis of the second coordinate system. We change the axes in the $(2, 3)$ coordinate plane as indicated. The quadratic form $F(r)$ then assumes the form

$$F(r) = c_{11} \xi_1^2 + c_{22} \xi_2^2 + c_{33} \xi_3^2. \tag{56}$$

That is, there is a coordinate system in space with the original origin in which a given quadratic form $F(x_1, x_2, x_3)$ admits the form (56). The form (56) is called *the canonical form of the quadratic form F*.

Considerations very like those set forth in §10 now show that every non-degenerate surface of order 2 has one of the five forms already listed in the Supplement to Chapter II: again see formulas (83)–(87).

Degenerate surfaces of order 2 are cylinders and cones. They are simple to describe. A cylinder of order 2 has the form

$$F(x_1, x_2) = 0 \tag{57}$$

in an appropriately chosen coordinate system. Here $F(x_1, x_2)$ is a polynomial of degree 2 in the variables x_1, x_2. Thus the equation

$$F(x_1, x_2) = 0$$

defines a curve of order 2 in the $(1, 2)$ coordinate plane. That is, the directrix of the cylinder (57) is a conic section or two intersecting lines or a single line or a single point or is no curve at all.

A cone of order two has the equation

$$F(x_1, x_2, x_3) = 0$$

in an appropriate coordinate system, where $F(x_1, x_2, x_3)$ is a quadratic form. Suppose that $F(x_1, x_2, x_3)$ has been reduced to the canonical form (56). Thus a cone of order 2 can be described by an equation

$$c_{11} x_1^2 + c_{22} x_2^2 + c_{33} x_3^2 = 0. \tag{58}$$

If all of the numbers c_{11}, c_{22}, c_{33} are different from 0 and have the same sign, then (58) is satisfied only by the origin. If one or two of the coefficients is zero, then we have a cylinder. If all three are different from zero and do not have the same sign, we may suppose that $c_{33} > 0$, $c_{11} < 0$, $c_{22} < 0$. The equation (58) can then be written in the form

$$x_3^2 = \frac{x_1^2}{a_1^2} + \frac{x_2^2}{a_2^2}.$$

This is the equation of a cone having elliptical intersection with each plane $x_3 = c$ $(c \neq 0)$.

This completes the classification of surfaces of order 2.

3. Conic Sections Revisited. Suppose that we have a Cartesian coordinate system in space, as for example in the Supplement to Chapter I. We can describe curves and surfaces in space in parametric form, by analogy with what we did with curves in the plane in §6. Let $x(t)$ be a vector function (in space) of a real parameter t:

$$x(t) = (x_1(t), x_2(t), x_2(t)).$$

The equation

$$x = x(t)$$

is then a parametric equation for a curve in space. That is, as t varies, the point $x(t)$ traces out a certain curve in space. In much the same way, suppose that $x(t_1, t_2)$ is a vector function (again in space) of two independent real parameters t_1 and t_2:

$$x(t_1, t_2) = (x_1(t_1, t_2), x_2(t_1, t_2), x_3(t_1, t_2)).$$

The equation

$$x = x(t_1, t_2)$$

defines a certain surface in space. As t_1 and t_2 take on all values, the point $x(t_1, t_2)$ traces out a certain surface S in space. We can regard t_1 and t_2 as "coordinates" of a point $x(t_1, t_2)$ on the surface S. Such parametric representations of surfaces are of great importance in geometry. Here we are concerned only with parametric representations for straight lines and planes.

Let M be a straight line in space. Let x_0 be a fixed point on M and let e be a vector of length 1 with initial point the origin 0 and parallel to the line M. The equation

$$x = x_0 + t e \tag{59}$$

is a parametric equation of the line M. As the parameter t runs through all real values, the point x given by equation (59) sweeps out the entire line M. Equation (59) has the same form as equation (45) of Chapter II and is derived in exactly the same way.

Now let Q be a plane in space. Let x_0 be any fixed point in Q and let e_1 and e_2 be two vectors of length 1 having initial point o, perpendicular to each other, and with the property that the plane containing o, e_1, and e_2 is parallel to the plane Q. Then the equation

$$x = x_0 + t_1 e_1 + t_2 e_2 \tag{60}$$

is a parametric equation for the plane Q. As t_1 and t_2 run through all real values, the point x defined by equation (60) will sweep out the entire plane Q. We may regard t_1 and t_2 as coordinates of the point x defined by (60), in a certain Cartesian coordinate system for the plane Q. This coordinate system in Q is obtained as follows. Draw a line M_1 in Q passing through the point x_0 and parallel to the vector e_1. Choose the direction on M_1 that coincides with the direction of e_1. Draw a line M_2 in Q going through x_0 and having the same relation to e_2 that M_1 has to e_1. The lines M_1 and M_2 are coordinate axes on Q, and t_1 and t_2 are the coordinates in this coordinate system of the point x defined by (60). Consider the line R in space that passes through 0 and is perpendicular to the plane Q. If the point x_0 on Q lies on the line R, then we can write x_0 as $a e_3$, where $a \geq 0$ and the vector e_3 has length 1 and is perpendicular to both e_1 and e_2. In this case, the vector equation (60) has the form

$$x = t_1 e_1 + t_2 e_2 + a e_3. \tag{61}$$

To recast the vector equation (61) in scalar form, we write the vectors e_1, e_2, e_3 in coordinate form:

$$e_i = (\alpha_{i,1}, \alpha_{i,2}, \alpha_{i,3}) \quad (i = 1, 2, 3).$$

The single vector equation (61) is equivalent to the three following scalar equations:

$$x_i = \alpha_{i,1} t_1 + \alpha_{i,2} t_2 + \alpha_{i,3} a \quad (i = 1, 2, 3). \tag{62}$$

Now let $F(x_1, x_2, x_3)$ be a function of three variables. The equation

$$F(x_1, x_2, x_3) = 0 \tag{63}$$

defines a certain surface in space. To find the intersection L of this surface, S, with the plane Q defined above, we must replace x_1, x_2, x_3 in (63) by the values written in (62). This gives us a function $G(t_1, t_2)$, and the curve L in the plane Q is defined by the equation

$$G(t_1, t_2) = 0.$$

Let us now choose $F(x_1, x_2, x_3)$ to be the equation of a circular cone in space (see para. 3 in the Supplement to Chapter II). One such equation is

$$x_1^2 + x_2^2 - k^2 x_3^2 = 0. \tag{64}$$

For nonzero k, (64) is the equation of a circular cone whose axis is the third coordinate axis and whose vertex is the origin. We choose the plane Q also in a special way, setting

$$e_1 = (1, 0, 0),$$
$$e_2 = (0, -\sin \gamma, \cos \gamma),$$
$$e_3 = (0, \cos \gamma, \sin \gamma),$$
$$a > 0.$$

The parametric equations (62) of the plane Q in this case are:

$$\left. \begin{array}{l} x_1 = t_1, \\ x_2 = -t_2 \sin \gamma + a \cos \gamma, \\ x_3 = t_2 \cos \gamma + a \sin \gamma. \end{array} \right\} \tag{65}$$

Substituting the values in (65) in the equation (64), we obtain

$$t_1^2 + (\sin^2 \gamma - k^2 \cos^2 \gamma) t_2^2 - 2a \sin \gamma \cos \gamma (1 + k^2) t_2$$
$$+ (\cos^2 \gamma - k^2 \sin^2 \gamma) a^2 = 0. \tag{66}$$

This is the equation of the curve L in the coordinates t_1, t_2 in the plane Q. Plainly the equation (66) defines a curve of order 2 in the plane Q. We leave it to the reader to determine the values of $a, k,$ and γ that yield ellipses, hyperbolas, and parabolas.

Thus the method of coordinates gives us the results of §11 almost automatically, while in §11 we obtained our results by subtle geometric constructions. This demonstrates the power of the method of coordinates.

Part II
Analysis of the Infinitely Small

Introduction to Part II

The principal concepts of mathematical analysis are derivatives and integrals. Before attaining its present day form, analysis was put together and developed over the course of many centuries, indeed one may say millennia. In any case, the works of Archimedes (287–212 B.C.) contain constructions that are now recognized as rudimentary forms of integrals and derivatives. Archimedes computed the area of a plane region bounded by an arc of a parabola and a straight line intersecting the parabola. He did this by the method of exhaustion, which involved inscribing a sequence of polygons in his figure. The areas of the polygons gradually exhaust the entire area of the figure. From our present point of view, this construction is a rudimentary sort of integration. So far as the method of exhaustion goes, Archimedes also had predecessors. Archimedes also constructed tangent lines to spirals, and from our present point of view this amounted to finding a derivative in a certain case.

The differential and integral calculus was discovered at the end of the 17th and beginning of the 18th centuries by Leibniz (1646–1716) and Newton (1642–1727). These scientists gave the principal algorithmic methods for both disciplines. They made their discoveries almost simultaneously and quite independently of each other. Neither of them was in any hurry to publish their results, since neither of them was able to perceive with sufficient clarity the nature of their own work. The algorithms that they discovered worked well and gave reliable results, which were important for applications. However, they did not understand clearly the basis of their algorithms. A certain atmosphere of mysticism lay over their thought. It is almost as if they followed the rule "Act and the truth will come later." The point was that in defining velocity, which is the derivative with respect to time, they used the totally unclear concept of an instantaneous interval of time. This mysterious instantaneous interval of time was perfectly small. Indeed in some calculations it was to be replaced by zero, while in others this replacement was not allowed. Velocity was defined as the ratio of the distance traveled in one moment of time to the duration of this moment. Plainly the duration of the moment cannot be taken as zero in such a calculation, since then the velocity would have the meaningless form $\frac{0}{0}$. Suppose that, like Newton, we write the duration of the moment by the single letter O. Then, calculating the distance traveled divided by the duration of the moment, one then has to replace the symbol O by zero. This in fact leads to the correct result. Leibniz encountered difficulties of just the same kind. Cauchy (1789–1857) found the way out of this confusion with the notion of a limit. The essence of this idea is as follows. The interval of time is not taken to be instantaneous, but has a quite definite finite length. Compute the distance traveled in this finite time to the time itself. In this well-defined ratio, let the interval of time begin to shrink, going to zero. One then finds what limit this ratio tends to. This limit is the velocity, or derivative.

As the differential and integral calculus developed, a certain new concept arose and assumed a very large rôle. This is the notion of an infinite series, which is to say the sum of an infinite number of summands. We write this in the form

$$z_1 + z_2 + \cdots + z_n + \cdots. \tag{1}$$

The use of such sums led as a rule to correct and important results, but occasionally to blunders. One must understand under what circumstances one may use infinite sums without hazard and what numerical values one is to give to sums (1). Here is an example of an infinite series that leads to a paradox:

$$1 - 1 + 1 - 1 + \cdots. \tag{2}$$

If we approach this sum as one does finite sums, we can group every term standing in an odd place with the following term in an even place. These summands cancel each other and so we would obtain 0 for the sum of the series (2). We can, however, group the terms differently. First set aside the first term, 1. Then group the second with the third, the fourth with the fifth, and so on. These grouped terms cancel each other, and so we would obtain 1 for the sum of the series (2). Some mathematicians adopted the point of view that the sum of (2) should be $\frac{1}{2}$. The following argument was used to support this view. Suppose that we assign a sum s to the series (2). Then, regrouping, we see that $s = 1 - s$, and so we find that $s = \frac{1}{2}$. Cauchy also set the theory of infinite series on a firm foundation. He defined precisely the conditions under which one can define the sum (1) and also what numerical value one should assign to it. He considered partial sums

$$s_n = z_1 + z_2 + \cdots + z_n,$$

for every positive integer n. This gives an infinite sequence of partial sums

$$s_1, s_2, \ldots, s_n, \ldots.$$

If this sequence of numbers converges to a limit s, then one says that the series (1) converges and that its sum is the number s.

The series of Taylor (1685–1731) underlines the rôle of infinite series in mathematical analysis. Taylor discovered the series that bear his name by studying a finite difference formula of Newton. The immense significance of

Taylor's series lies in the fact that all functions that were studied in his time can be written as Taylor series. There thus arose an impression that every function can, at least in pieces, be represented by a Taylor series. This impression proved to be wholly false, as was proved much later.

A new direction in analysis arose with the study of complex-valued functions of a complex variable. Many points of confusion arose in the study of these functions. However, with the participation of such great mathematicians as Euler (1707–1783) and Gauss (1777–1855), enough facts about such functions were accumulated that a coherent theory of analytic functions of a complex variable could be formulated. Here again decisive results were due to Cauchy. He proved that the integral of a complex function of a complex variable along a closed curve is zero, provided that there are no jumps or singularities of the function in the interior of the curve. Cauchy did not mention that the function in question must have a derivative and indeed a continuous derivative, but he used this hypothesis in carrying out his proof. It is hard to conceive that such an omission is a mistake. Either Cauchy thought that differentiability of the function is so natural an hypothesis that it need not be mentioned, or he thought that all continuous functions admit derivatives. Cauchy derived his famous integral formula from this theorem. His integral formula allows one to compute the values of an analytic function inside a closed curve from its values on the curve itself. With these results, Cauchy founded the theory of functions of a complex variable, a theory that plays a great rôle in contemporary analysis and finds an immense number of applications.

Chapter IV
Series

In mathematics one very often must consider sums with an infinite number of summands, or, as they are called, *infinite series*. Such a sum or series is written in the form

$$z_1 + z_2 + \cdots + z_n + \cdots ,\qquad (1)$$

where $z_1, z_2, \ldots, z_n, \ldots$ are real or complex numbers. In elementary algebra, a special case of such a series is frequently studied:

$$a + a q + a q^2 + \cdots + a q^n + \cdots ,$$

where $|q| < 1$. It is established that such a geometric progression has a sum and that it is equal to

$$\frac{a}{1-q}.$$

The question arises as to what conditions one must impose on the numbers z_n in order for the series (1) to have a sum. That not everything is clear in this connection is shown by the following example. Consider the series

$$1 - 1 + 1 - 1 + \cdots \qquad (2)$$

((2) in the preceding Introduction to Part II). As noted there, both 0 and 1 are equally reasonable "sums" for (2). If we take the partial sums $1, 0, 1, \ldots$ of (2) and *average* them, we get the sequence of numbers $1, \frac{1}{2}, \frac{2}{3}, \frac{1}{2}, \frac{3}{5}, \ldots$ with

$$a_n = \tfrac{1}{2} \quad \text{for } n \text{ even and } \quad a_n = \tfrac{1}{2} + \frac{1}{2n} \quad \text{for } n \text{ odd.}$$

For all large enough indices n, the numbers a_n will be as close as you please to the number $\frac{1}{2}$. This might also justify us in assigning $\frac{1}{2}$ as the "sum" of the series (2). In any case, the example (2) shows that we must approach infinite series with caution: we cannot operate them as casually as we do with finite sums. We must first define exactly what we mean by the sum of the series (1). For this purpose, we consider *partial sums of the series* (1), which are defined as

$$s_n = z_1 + z_2 + \cdots + z_n.$$

The sum s_n depends of course on the index n. We may consider its behavior as n grows larger and larger. If the numbers s_n converge to a fixed limit s as n grows without bound, we say that the series (1) *converges to the sum s*. However, convergent series may display peculiarities that do not arise for finite sums. It may occur, for example, that the terms of a convergent series (1) can be rearranged so that the series again converges but converges to a number different from the sum of the original series. To avoid this pathology, as well as some others, we single out a special class of convergent series. The series (1) is said *to converge absolutely* if the absolute values of its terms form a convergent series, that is, if the series

$$|z_1|+|z_2|+ \cdots +|z_n|+ \cdots$$

converges. One can prove that the sum of an absolutely convergent series is independent of the order in which we take its summands, that two absolutely convergent series can be multiplied by the ordinary rules for multiplying finite sums, and so on. That is, one can calculate quite freely with absolutely convergent series and obtain correct results.

Series are important in particular because they can be used to compute various numbers that are important in mathematics. For example, we have the well-known formula

$$\tfrac{1}{4}\pi=1-\tfrac{1}{3}+\tfrac{1}{5}- \cdots +\frac{(-1)^{n+1}}{2n-1}+ \cdots . \tag{3}$$

By using the series (3), one can compute the number π to any degree of accuracy. One should note, however, that this is not the best way to compute π, since the series (3) converges very slowly.

The computation of functions by means of series is no less important. Suppose that the terms of the series (1) are functions of a real or complex variable z. Then, if the series converges, its sum too is a certain function $f(z)$. We write accordingly

$$f(z)=z_1(z)+z_2(z)+ \cdots +z_n(z)+ \cdots . \tag{4}$$

In considering series of the form (4), there are two distinct possibilities. We may have a known function $f(z)$, and look for an expansion of $f(z)$ in the form (4). On the other hand, we may wish to define a function $f(z)$ from its expansion (4). Starting with an expansion (4), we may be able to study various important properties of the function $f(z)$.

Of particular importance are series (4) in which $z_n(z)$ has the form $a_n z^n$ for constants a_n:

$$a_0+a_1 z+a_2 z^2+ \cdots +a_n z^n+ \cdots . \tag{5}$$

Such series are called *power series*. The variable z may be either real or complex, and the same is true of the constants a_n. It was discovered centuries ago that many important functions can be written as power series. At one time mathematicians harbored the view that all functions can be expanded, at least in pieces, in power series. This viewpoint, however, has turned out to be quite false.

In §20, the last section of the present chapter, we make a careful study of functions of a complex variable z that can be written as power series (5). The most interesting property that we establish in §20 is the following. Plainly every series (5) converges for $z=0$. Possibly the a_n's grow so rapidly in absolute value that the series diverges for all other z's. We may exclude this trivial case. Then the series (5) admits what is called its *radius of convergence*. This is a positive number r with the property that the series (5) converges (indeed absolutely) for all z such that $|z|<r$, and diverges for all z such that $|z|>r$. That is, the series (5) defines a function $f(z)$ for all z such that $|z|<r$. The case $r=\infty$ is not excluded. For $r=\infty$, the series (5) converges and so defines a function $f(z)$ defined in the entire complex plane.

In §18, we take up a function defined by a power series that is extremely important in analysis. This is the function $\exp(z)$ of the complex variable z, defined by the power series

$$\exp(z)=1+\frac{z}{1}+\frac{z^2}{1\cdot 2}+\cdots+\frac{z^n}{n!}+\cdots. \qquad (6)$$

The power series in (6) converges for all complex numbers z, and so the function $\exp(z)$ is defined in the entire plane of the complex variable z. We will establish in §18 a number of properties of $\exp(z)$. Thus for real values $z=x$, we obtain

$$\exp(x)=e^x, \qquad (7)$$

while for pure imaginaries $z=iy$ we have the identity

$$\exp(iy)=\cos y+i\sin y. \qquad (8)$$

From the identities (6), (7), and (8) we immediately obtain power series expansions of the fundamental transcendental functions e^x, $\cos y$, and $\sin y$ for real values of their arguments.

As has already been remarked, we encounter two distinct points of view regarding power series expansions (5). First, we can suppose that we know the function $f(z)$ from other considerations, and then try to show that it admits a power series expansion (5). That is, we must show that for a given value of z, the power series converges to the (known) number $f(z)$. Second, we can define the function $f(z)$ from its power series expansion (5). To show that this procedure is legitimate, we have to prove that the power series in (5) converges. To put this notion on a firm foundation, we have to have a clear idea of what real and complex numbers are and also obtain a criterion for convergence of a sequence of numbers. Sections 1–4 of the present chapter are devoted to these questions.

§12. Convergent Sequences of Numbers

Convergent sequences of numbers play an extremely important rôle in mathematics. Let

$$s_1, s_2, \ldots, s_n, \ldots \tag{1}$$

be a sequence of real numbers. We say that this sequence *converges to the number s* or *has limit s* provided that the numbers s_n are arbitrarily close to the number s for all sufficiently large values of the index n. We write

$$\lim_{n \to \infty} s_n = s. \tag{2}$$

Suppose that $s = 0$, i.e., that the sequence (1) converges to 0. The quantity s_n, which is a function of the positive integral variable n, is sometimes in this case called *infinitely small* or an *infinitesimal*. It would be more precise to say that the sequence is *infinitely decreasing*, since one is really considering the convergence to 0 of the numbers s_n as n increases.

It is plain that (2) is equivalent to

$$\lim_{n \to \infty} (s_n - s) = 0. \tag{3}$$

In other terms, the sequence (1) converges to s if and only if the sequence

$$s_1' = s_1 - s, \qquad s_2' = s_2 - s, \ldots, s_n' = s_n - s, \ldots$$

converges to zero. If (3) holds, then the quantity s_n' is an infinitesimal. Thus a study of the process of convergence leads for the most part to a study of infinitesimals.

Decimal Expansions. Let us illustrate the process of convergence with a familiar example. Consider infinite decimal expansions, which are familiar to us from elementary school arithmetic.

Let us write the infinite decimal expansion

$$0.999\ldots. \tag{4}$$

After the decimal point we think of an infinite number of occurrences of the number 9. It is well known from arithmetic that this decimal is equal to 1. The precise mathematical sense of this assertion is based on a consideration of the infinite sequence

$$s_1 = 0.9, \qquad s_2 = 0.99, \ldots, s_n, \ldots, \tag{5}$$

where s_n is the finite decimal $0.999\ldots9$ having n occurrences of the number 9. It is easy to show that for the sequence (5), we have

$$s_n' = s_n - 1 = -\frac{1}{10^n}.$$

Using the usual decimal representation of positive integers, we can show with no difficulty that $-\dfrac{1}{10^n}$ is an infinitesimal, which is to say, the sequence (5) converges to the limit 1. This is the correct interpretation of the assertion that the infinite decimal (4) is equal to 1.

Consider a second infinite decimal:

$$0.333\ldots. \tag{6}$$

In (6), after the decimal point we have an infinite number of occurrences of the number 3. We obtain this infinite decimal when we carry out ordinary division on the fraction $\frac{1}{3}$. Carrying out this division over and over, we obtain the sequence

$$s_1 = 0.3, \qquad s_2 = 0.33, \ldots, s_n, \ldots. \tag{7}$$

The number s_n here is the finite decimal with exactly n occurrences of the number 3 after the decimal point.

It is again easy to show that for the sequence (7), the equalities

$$s'_n = s_n - \tfrac{1}{3} = -\frac{1}{3 \cdot 10^n}$$

hold. The quantity $s'_n = -\dfrac{1}{3 \cdot 10^n}$ is obviously an infinitesimal, and so the sequence (7) converges to the rational number $\frac{1}{3}$. This is the exact meaning of the assertion that the infinite decimal (6) is equal to $\frac{1}{3}$.

The infinite decimals (4) and (6) are completely specific and are also periodic. We turn now the the case of a general infinite decimal, which we write in the form

$$0.f_1 f_2 f_3 \ldots. \tag{8}$$

In the expression (8), the numbers $f_1, f_2, f_3, \ldots$ that follow the decimal point can be quite arbitrary, so long as each of them is one of the numbers 0, 1, 2, 3, 4, 5, 6, 7, 8, or 9. Let us write down a sequence of finite decimals corresponding to the infinite decimal (8):

$$s. = 0.f_1, \qquad s_2 = 0.f_1 f_2, \ldots, s_n, \ldots. \tag{9}$$

In (9), the number s_n is the finite decimal in which after the zero and the decimal point there are the n numbers $f_1, f_2, \ldots, f_n$. It is known that if the infinite decimal (8) is ultimately periodic, then the sequence of numbers (9) converges to a certain rational number s. Indeed, we have

$$|s'_n| = |s_n - s| \leqq \frac{1}{10^n}.$$

Thus s'_n is an infinitesimal. This is what we mean by saying that the infinite decimal (8) is equal to the rational number s. On the other hand, suppose that the infinite decimal (8) is not ultimately periodic. Then there is no rational number s to which the sequence of numbers (9) converges. It is thus reasonable in this case to ask if we can think of the infinite decimal (8) as a number. There is no doubt that a need for such numbers exists. Even in elementary geometry we encounter curves whose lengths are represented by infinite nonperiodic decimals. The following examples are classical. 1) Consider the length of the hypotenuse of an isosceles right triangle the legs of which have length 1. The length of this hypotenuse is $\sqrt{2}$. 2) Consider a circle of diameter 1. The length of this circle, often called its *perimeter*, is the number π. We can compute $\sqrt{2}$ by the standard method of extracting square roots. Doing this, we get an infinite nonperiodic decimal of the form (8). The sequence (9) for this decimal converges to $\sqrt{2}$ in a natural sense, namely, the sequence of squares

$$s_1^2, s_2^2, \ldots, s_n^2, \ldots$$

converges to the number 2.

The state of affairs for the number π is similar. These two examples show that the requirements of elementary geometry cannot be satisfied by rational numbers alone. We are compelled to admit the existence of irrational numbers, which is to say, numbers that are not rational. One can of course identify not merely two irrational numbers that are needed in mathematics, but an infinite number of them. We are thus faced with the problem of describing all of the numbers that are required in mathematics. The key to solving this problem lies in the theory of convergent sequences of numbers. We will take up this theory in §14.

A Formal Definition of Convergence. The foregoing discussion gives a fairly complete, though intuitive, description of convergence. We will now give the formal definition, which is generally agreed upon.

Definition. Let

$$s_1, s_2, \ldots, s_n, \ldots \tag{10}$$

be a sequence of real numbers. We say that this sequence *converges to the number s* if, for every positive number ε, there exists a natural number v that depends upon ε and has the property that for all $n > v$, the inequality

$$|s_n - s| < \varepsilon \tag{11}$$

holds.

To make the process of convergence clear, a geometric interpretation is useful. Let us represent each of the numbers s_n in the sequence (10) by a point on the axis of abscissas, namely, the point whose abscissa is equal to the number s_n. We represent the number s in like manner. Let U_ε denote the set of all points x on the axis of abscissas such that the inequality

$$|x-s|<\varepsilon \qquad (12)$$

holds. Here x denotes not only a point but also the value of its abscissa. The set U_ε is an interval on the axis of abscissas, and we call is *the ε-neighborhood of the point s*. The definition of convergence means exactly that the sequence of points (10) converges to the point s if and only if for every positive number ε, the neighborhood U_ε contains all points of the sequence (10) with the possible exception of a finite number of these points.

The proof of this assertion offers no difficulties, and I will not give it here. In somewhat picturesque language, one may say that the points of the sequence (10) are collected mainly at the point s.

We return to the simple examples (4) and (6). Here the numbers $s_1, s_2, s_3, \ldots$ are increasing and are all less than their limits. This of course is not obligatory. The numbers of the sequence (10) can be partly greater than s and partly less than s, and of course they can partly coincide with s. It may also happen that numbers s_n with different indices are in fact equal. An extreme case of this occurs when all of the numbers s_n are the same number. Here we do not have a gradual approximation to s by the numbers s_n: all of them are already at their limit s. Nevertheless, even under these conditions the sequence (10) converges to s.

The Case of Complex Numbers. So far we have considered convergence only for sequences of real numbers. The notion of convergence of sequences of complex numbers is also important. The case of complex numbers is reduced at once to the case of real numbers, and it contains nothing that is essentially novel.

Consider a sequence

$$r_1 = s_1 + i t_1, \quad r_2 = s_2 + i t_2, \ldots, r_n = s_n + i t_n, \ldots \qquad (13)$$

of complex numbers. Suppose that there is a complex number r for which the sequence of real numbers

$$|r_1 - r|, \quad |r_2 - r|, \ldots, |r_n - r|, \ldots$$

converges to zero, that is, such that

$$\lim_{n \to \infty} |r_n - r| = 0. \qquad (14)$$

We then say that the sequence (13) converges to r, or has limit r, and we write

$$\lim_{n \to \infty} r_n = r = s + i t. \qquad (15)$$

That is, we define (15) as holding if and only if (14) holds.

Suppose that all of the complex numbers appearing in (13) are actually real. Then our definition of convergence coincides with the definition for sequences of real numbers.

If the limit r of the sequence (13) is 0, we say that the complex numbers r_n comprise *an infinitesimal*, or more precisely, is *infinitely decreasing*. Plainly the relation (15) holds if and only if the quantity

$$r_n' = r_n - r$$

is an infinitesimal.

It is also useful to know that (15) is equivalent to the two relations

$$\lim_{n \to \infty} s_n = s \quad \text{and} \quad \lim_{n \to \infty} t_n = t.$$

The proof is simple, and I omit it. We can interpret (15) in geometric terms, which give it a useful pictorial form. Represent the points r_n of the sequence (13) and the point r as points in the complex plane. For a positive number ε, let U_ε be the disk in the complex plane with center r and radius ε, which is to say, U_ε is the set of all complex numbers z such that

$$|z - r| < \varepsilon. \tag{16}$$

The disk U_ε is called *the ε-neighborhood of the point r* in the complex plane. Plainly the sequence (13) converges to r if and only if, for every positive number ε, the neighborhood U_ε contains all but a finite number of the points in the sequence (13).

Subsequences. Consider a sequence

$$r_1, r_2, \ldots, r_n, \ldots \tag{17}$$

of real or complex numbers. Choose a certain infinite part of its members and arrange these selected members in the same order that they have in the original sequence (17). We obtain a new sequence, which is called *a subsequence of the original sequence* (17). Such a subsequence can always be written in the form

$$r_{n_1}, r_{n_2}, \ldots, r_{n_k}, \ldots, \tag{18}$$

where the indices

$$n_1, n_2, \ldots, n_k, \ldots$$

are a certain sequence of natural numbers, arranged in increasing order. Frequently it is necessary to denote the terms of the sequence (18) by new letters with new indices so that the new indices are $1, 2, 3, \ldots, k, \ldots$. For example, we can write

$$r_1' = r_{n_1}, r_2' = r_{n_2}, \ldots, r_k' = r_{n_k}, \ldots.$$

It is obvious that if the original sequence (17) converges to a number r, then all of its subsequences (18) also converge to the number r. That is, if

$$\lim_{n \to \infty} r_n = r,$$

then we also have

$$\lim_{k \to \infty} r_{n_k} = \lim_{k \to \infty} r_k' = r.$$

On the other hand, it frequently happens that the original sequence (17) diverges, but that one can find a convergent subsequence. This phenomenon plays a significant rôle in our later work.

The Main Rules of the Theory of Limits. Let us now set down some simple properties of infinitely small quantities. We will deal with complex quantities, remembering always that real quantities are special complex quantities. Let us write a sequence of complex numbers:

$$c_1, c_2, \ldots, c_n, \ldots.$$

We call this sequence *bounded* if there is a positive number c such that

$$|c_n| \leq c \tag{19}$$

for all indices n. If (19) holds and r_n is an infinitely small quantity, that is, if

$$\lim_{n \to \infty} r_n = 0,$$

the quantity $c_n r_n$ is also infinitely small: that is,

$$\lim_{n \to \infty} c_n r_n = 0.$$

The proof of this fact is simple. We note as well that the sum and difference of infinitely small quantities is infinitely small.

Using these simple remarks about infinitely small quantities, we can establish the five principal rules of the theory of limits.

Let us write two sequences of complex numbers:

$$z_1, z_2, \ldots, z_n, \ldots \tag{20}$$

and

$$w_1, w_2, \ldots, w_n, \ldots. \tag{21}$$

We suppose that the sequences (20) and (21) converge to numbers z and w respectively:

$$\lim_{n \to \infty} z_n = z \quad \text{and} \quad \lim_{n \to \infty} w_n = w. \tag{22}$$

Rule 1.

$$\lim_{n \to \infty} (z_n + w_n) = z + w,$$

or in words, the limit of a sum is equal to the sum of the limits.

Rule 2.

$$\lim_{n \to \infty} (z_n - w_n) = z - w,$$

or in words, the limit of a difference is equal to the difference of the limits.

Rule 3.

$$\lim_{n \to \infty} z_n w_n = z w,$$

or in words, the limit of a product is equal to the product of the limits.

Rule 4. Suppose that $w \neq 0$. Then we have

$$\lim_{n \to \infty} \frac{z_n}{w_n} = \frac{z}{w}, \tag{23}$$

or in words, the limit of a quotient is equal to the quotient of the limits, provided that the limit of the divisor is not zero.

Rule 5. Suppose that the quantities z_n and w_n are real and that

$$z_n \leqq w_n. \tag{24}$$

Then we have $z \leqq w$, that is, one may pass to the limit with the sign "less than or equal to". Note however that the inequalities $z_n < w_n$ do *not* imply that $z < w$.

The proofs of these rules are simple. I will give here only the proofs of (4) and (5), which are the most complicated.

We write

$$z_n = z + \hat{z}_n \quad \text{and} \quad w_n = w + \hat{w}_n, \tag{25}$$

so that $\hat{z}_n$ and $\hat{w}_n$ are infinitely small quantities. We form the differences

$$\frac{z_n}{w_n} - \frac{z}{w}. \tag{26}$$

Rewrite the first fraction by using the formulas (25). We obtain

$$\frac{z + \hat{z}_n}{w + \hat{w}_n} - \frac{z}{w} = \frac{z w + \hat{z}_n w - z w - z \hat{w}_n}{(w + \hat{w}_n) w} = \frac{\hat{z}_n w - z \hat{w}_n}{(w + \hat{w}_n) w}.$$

The numerator of the last fraction written is an infinitely small quantity, and the quantity $\dfrac{1}{(w+\hat{w}_n)\,w}$ is bounded for sufficiently large n, since $w \neq 0$. Therefore the difference (26) is an infinitely small quantity. This proves the equality (23).

We now prove rule 5. Using the notation in (25), we write (24) in the form

$$w_n - z_n = (w + \hat{w}_n) - (z + \hat{z}_n) \geqq 0,$$

or equivalently

$$(w - z) + (\hat{w}_n - \hat{z}_n) \geqq 0. \tag{27}$$

Since the quantity $\hat{w}_n - \hat{z}_n$ is infinitely small, we infer from (27) that the number $w - z$ cannot be negative. In fact, if $w - z$ were negative, $w_n - z_n$ would be negative for all large enough n, in view of the fact that $\hat{w}_n - \hat{z}_n$ is infinitely small.

In the three following examples, we present some notions that will be used only farther on. It is therefore recommended that they be studied only when the reader encounters references to them.

Example 1. The reader is familiar with the notion of a function from Chapter II, §4. We recall the definition. A quantity w is said to be a function of the quantity z if there is a rule which enables us to compute w once we know z. We write $w = f(z)$. A polynomial

$$w = f(z) = a_0 z^k + a_1 z^{k-1} + \cdots + a_k$$

(k is a natural number) is an example of a function. What are called *continuous functions* are very important in mathematical analysis. A great deal of attention is lavished on them in textbooks on analysis. Not wishing to lay a huge amount of stress on continuous functions, I give the definition of continuous function only in this example. I treat the real and complex cases together.

Definition. A function $f(z)$ is said to be *continuous at a point* $z = z_0$ if, for every sequence

$$z_1, z_2, \ldots, z_n, \ldots$$

of points where $f(z)$ is defined, the relation

$$\lim_{n \to \infty} z_n = z_0 \tag{28}$$

implies that

$$\lim_{n \to \infty} f(z_n) = f(z_0). \tag{29}$$

The function $f(z)$ is said simply to be *continuous* if it is continuous at every point where it is defined.

From rules 1–4 of the theory of limits, it is obvious that if $f(z)$ and $g(z)$ are continuous at a point $z=z_0$ where they are defined, then the functions

$$f(z)+g(z), \quad f(z)-g(z), \quad f(z)\,g(z), \quad \text{and} \quad \frac{f(z)}{g(z)} \tag{30}$$

are also continuous at $z=z_0$. The last-written function in (30) is of course defined at $z=z_0$ only if $g(z_0)\neq0$. For every positive integer k, the function $[f(z)]^k$ is continuous at $z=z_0$ if $f(z)$ is.

The functions $f(z)=z$ and $f(z)=c$ (c a constant) are obviously continuous. It follows that every polynomial

$$\varphi(z)=a_0 z^k+a_1 z^{k-1}+\cdots+a_k$$

is a continuous function. Also, if $\psi(z)$ is a second polynomial, the rational function $\dfrac{\varphi(z)}{\psi(z)}$ is continuous at every point where ψ does not vanish.

Example 2. Let us give some examples of discontinuous functions. We define a real-valued function $f(x)$ of a real variable x by the rules

$$f(x)=1 \quad \text{for } x>0,$$
$$f(x)=-1 \quad \text{for } x<0. \tag{31}$$

(For the graph of this function, see Fig. 58.) We have not defined the function $f(x)$ for $x=0$. If we do not define f at 0, but only for positive x and negative x, our definition of continuity shows that $f(x)$ is continuous everywhere. This is easy to check. Now we ask, is it possible to define $f(x)$ for $x=0$ in such a way that the resulting function, which is defined for all real values of x, is continuous? Plainly this is impossible. Indeed, suppose that $x_1,x_2,\ldots,x_n,\ldots$ is a sequence of positive numbers that converges to 0. Then we have $f(x_n)=1$ for all n, and from our definition of continuity, we see that we must define $f(0)$ as 1. If we take a sequence of negative numbers converging to 0, we see in the same way that we must define $f(0)$ as -1. Therefore there is no way to define $f(0)$ so that $f(x)$ is continuous at $x=0$.

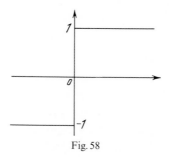

Fig. 58

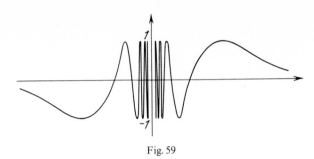

Fig. 59

We now consider the real-valued function $\sin\left(\dfrac{1}{x}\right)$, defined for all non-zero real numbers x. We take as known the fact that the function $\sin\varphi$ is continuous. From this it easily follows that the function $\sin\left(\dfrac{1}{x}\right)$, defined only for nonzero real numbers x, is also continuous. It is quite impossible to define this function at 0 in such a way that it is continuous at 0. To convince one's self of this, it is useful to contemplate the graph of the function (see Fig. 59).

Example 3. We consider here complex numbers, identifying them as usual with points of the complex plane. We have already defined the ε-neighborhood U_ε of a point r of the complex plane (see (16) and the preceding discussion). Note that we do not include the boundary points in defining the disk U_ε. Let M be a certain infinite set of points of the complex plane. A point r is called *a limit point of M* if every neighborhood of r contains an infinite number of points of M. If M contains only real numbers, it is clear that any limit point of M is a real number. If $r = s + it$ with $t \neq 0$, then the neighborhood $U_{|t|}$ of r contains no real numbers and so r cannot be a limit point of any set contained in the real axis.

Now let r be a limit point of a set M. We can choose a sequence

$$r_1, r_2, \ldots, r_n, \ldots \tag{32}$$

of points in M all of which are distinct and which converges to r. To construct the sequence (32), consider the neighborhood $U_{1/n}$ of the point r, for every positive integer n. Let r_1 be any point in M and in U_1 that is different from r. Suppose that $r_1, r_2, \ldots, r_n$ have been chosen. Then let r_{n+1} be any point in M and in $U_{1/(n+1)}$ that is different from r and also different from all of the points $r_1, r_2, \ldots, r_n$. Since there are an infinite number of points in M that are in $U_{1/(n+1)}$, r_{n+1} can always be found.

A set M is said to be *closed* if it contains all of its own limit points. For example, a line segment on the real axis, which consists of all real numbers x such that $a \leq x \leq b$, is a closed set.

§13. Infinitely Small Quantities

In the preceding section, we gave precise definitions of the limit of a sequence

$$r_1, r_2, \ldots, r_n, \ldots$$

of real or complex numbers, written in symbols as

$$\lim_{n \to \infty} r_n = r. \tag{1}$$

Formula (15) of §12 shows that we test to see whether or not (1) holds by looking at certain sequences of nonnegative real numbers to see whether or not they have limit zero, or, as we say, are infinitely small. In the present section, we take up some typical and frequently encountered infinitely small quantities. It behooves us to become as skillful at recognizing infinitely small quantities as we are at using the multiplication table. Our principal technique in identifying infinitely small quantities is to compare them with previously known infinitely small quantities, which we have learned and fixed in our memories.

The Symbols O and o. A basic technique in comparing two sequences s_n and t_n is to study their ratio $\dfrac{t_n}{s_n}$ and to see what happens to this ratio as n increases without bound. There are two important cases here, which have been given special notation.

1. The quantity $\dfrac{t_n}{s_n}$ is bounded for all n. That is, there exists a constant c such that

$$\left| \frac{t_n}{s_n} \right| < c$$

for all n. In this case we write

$$t_n = O(s_n). \tag{2}$$

2. The quantity $\dfrac{t_n}{s_n}$ is infinitely small, which is to say that

$$\lim_{n \to \infty} \frac{t_n}{s_n} = 0.$$

In this case we write

$$t_n = o(s_n). \tag{3}$$

It is clear that the relation (3) implies the relation (2). For, if the ratio $\dfrac{t_n}{s_n}$ goes to 0 as $n \to \infty$, it must be bounded over all n.

The symbols O and o are very convenient. They are widely used in analysis, and one should keep their meaning clearly in mind.

If the quantity s_n is infinitely small and (2) holds, then the quantity t_n is also infinitely small. Suppose that a quantity s_n is infinitely small and that (3) holds. We may then say that the quantity t_n goes to zero faster than the quantity s_n. The expressions (2) and (3) are frequently used when the sequence $s_1, s_2, \ldots, s_n, \ldots$ consists of all 1's. We then write

$$t_n = O(1) \tag{4}$$

and

$$t_n = o(1). \tag{5}$$

The expression (4) means that the quantity t_n is bounded, and (5) that it is infinitely small.

We frequently compare a quantity t_n not with a quantity s_n but with a quantity s_n^k, where k is some natural number. Suppose that we have

$$t_n = (s_n^k),$$

s_n being an infinitely small quantity. We then say that the quantity t_n *has order of smallness k with respect to the quantity* s_n. Such orders of smallness also play an important part in analysis.

In analysis, we must consider not only infinitely small quantities but infinitely large ones. We say that the quantity s_n is *infinitely large* if the number $|s_n|$ grows unboundedly as n increases. That is, for every positive number c, there is to be a natural number v such that the inequality

$$|s_n| > c$$

holds for all $n > v$. If this occurs, we write

$$\lim_{n \to \infty} s_n = \infty. \tag{6}$$

In the expression (6), the symbol $\lim_{n \to \infty}$ has a different meaning from the case in which a finite number appears on the right side.

Sometimes it is useful to consider two mutually reciprocal quantities s_n and t_n:

$$s_n t_n = 1$$

for all n. The quantity t_n in this case is infinitely small if and only if the quantity s_n is infinitely large. That is to say, the relations

$$\lim_{n \to \infty} s_n = \infty \quad \text{and} \quad \lim_{n \to \infty} t_n = 0 \tag{7}$$

are equivalent if s_n and t_n are reciprocals.

The simplest example of an infinitely large quantity is $s_n = n$. From (7) we infer that $t_n = \dfrac{1}{n}$ is infinitely small, or in formulas

$$\lim_{n \to \infty} n = \infty, \quad \lim_{n \to \infty} \frac{1}{n} = 0.$$

Newton's Binomial Theorem. In cases more complicated than the simple cases $s_n = n$ and $t_n = \dfrac{1}{n}$, we test sequences for having limit 0 or ∞ by using one or another algebraic formula. The famous "binomial theorem" of Newton is one of these. Despite the fact that this theorem appears in elementary algebra courses, I will prove it here.

We first recall the notion

$$n! = 1 \cdot 2 \cdot 3 \cdot \, \cdots \cdot n.$$

That is, $n!$ (read n *factorial*) for a positive integer n is the product of all of the integers $1, 2, 3, \ldots, n$. It is convenient to define $0!$ by

$$0! = 1.$$

Newton's binomial theorem gives the nth power of a binomial $z + w$ as a polynomial in powers of z and w. This identity is

$$(z+w)^n = \sum_{\substack{p,q \geq 0 \\ p+q=n}} \frac{n!}{p!\,q!} z^p w^q. \qquad (8)$$

The symbol $\displaystyle\sum_{\substack{p,q \geq 0 \\ p+q=n}}$ means that we must take the sum of all terms to the right of the summation sign for nonnegative integers p and q such that $p+q=n$. We prove the identity (8) by induction on n. It is first clear that (8) holds for $n=1$:

$$(z+w)^1 = z + w = \frac{1!}{1!\,0!} z + \frac{1!}{0!\,1!} w.$$

We suppose now that the identity (8) holds for the value $n = k-1$. We will prove that then it also holds for $n=k$. Let us compute the product $(z+w)^{k-1}(z+w)$, using (8) for the first factor. The identity (8) for $n = k-1$ contains all expressions

$$\frac{(k-1)!}{i!\,j!} z^i w^j \qquad (9)$$

where i and j are nonnegative integers and $i+j = k-1$.

When we multiply (9) by z, we obtain a term containing $z^{i+1} w^j$. When we multiply (9) by w, we obtain a term containing $z^i w^{j+1}$. For these two monomials to be equal to say $z^p w^q$, we must have for the first

$$i = p-1 \quad \text{and} \quad j = q,$$

while for the second we must have

$$i = p \quad \text{and} \quad j = q-1.$$

Computing the coefficients in these two cases and adding them, we obtain

$$\frac{(k-1)!}{(p-1)!q!}+\frac{(k-1)!}{p!(q-1)!}.$$

In this sum, multiply the numerator and denominator of the first summand by p and the numerator and denominator the second summand by q. This gives us a common denominator, so that the sum just written is equal to

$$\frac{(k-1)!p+(k-1)!q}{p!q!}=\frac{(k-1)!(p+q)}{p!q!}=\frac{k!}{p!q!}.$$

Thus (8) also holds for $n=k$ if it holds for $n=k-1$. This completes our proof by induction.

Let us write the coefficient $\dfrac{n!}{p!q!}$ in a slightly different form, dividing numerator and denominator by $p!$. We obtain

$$(z+w)^n=z^n+\frac{n}{1!}z^{n-1}w+\frac{n(n-1)}{2!}z^{n-2}w^2+\cdots$$

$$+\frac{n(n-1)\cdots(n-q+1)}{q!}z^{n-q}w^q+\cdots+w^n. \tag{10}$$

This is the form in which Newton's binomial theorem is usually stated in elementary algebra.

Some Specific Infinitely Small Quantities. We now take up some important examples of positive quantities that are infinitely small and their reciprocals, which are infinitely large. The connection between the two classes is pointed out in (7).

The geometric progression

$$1,\alpha,\alpha^2,\alpha^3,\ldots,\alpha^n,\ldots, \tag{11}$$

where α is a positive number, is of primary importance. For $0<\alpha<1$, the sequence (11) has limit 0. For $\alpha>1$, the sequence (11) increases unboundedly. We write:

$$\lim_{n\to\infty}\alpha^n=0 \quad \text{for } \alpha<1 \tag{12}$$

and

$$\lim_{n\to\infty}\alpha^n=\infty \quad \text{for } \alpha>1. \tag{13}$$

We note first that (12) and (13) are equivalent. Write $\beta=\dfrac{1}{\alpha}$ and form the sequence

$$1,\beta,\beta^2,\ldots,\beta^n,\ldots. \tag{14}$$

The sequences (11) and (14) are mutually reciprocal, and so one of them has limit 0 if and only if the other has limit ∞. Note too that if one of α and β is less than 1, then the other is greater than 1. This shows that (12) and (13) are equivalent. Thus to prove (12) and (13), it suffices to prove (13). For $\alpha > 1$, we write

$$\alpha = 1 + \delta,$$

where δ is a positive number. From (10) we obtain

$$\alpha^n = (1 + \delta)^n = 1 + n\delta + \cdots,$$

where the terms indicated by $\cdots$ in the right side are positive. That is, we have proved that

$$\alpha^n > n\delta.$$

Since δ is positive, the quantity $n\delta$ increases unboundedly as n increases. The same is therefore true of α^n. This proves (13) and (12).

The infinitely small quantity α^n $(0 < \alpha < 1)$ serves frequently in analysis as a scale of comparison. Therefore one should keep (12) well in mind so as to be able to apply it when needed without long reflection.

The technique we used in proving (13) allows us to prove the stronger assertion

$$\lim_{n \to \infty} \frac{\alpha^n}{n^k} = \infty \tag{15}$$

for all $\alpha > 1$ and all positive integers k. For the reciprocal, we get

$$\lim_{n \to \infty} n^k \alpha^n = 0 \tag{16}$$

for all α such that $0 < \alpha < 1$ and all positive integers k. The relation (15) asserts that for $\alpha > 1$, the quantity α^n goes to infinity faster than any positive integral power n^k. The relation (16) asserts that for $0 < \alpha < 1$, the quantity α^n goes to zero faster than the quantity $\frac{1}{n^k}$, where k is an arbitrary positive integer.

It suffices to prove only (16), since (15) is an immediate consequence. We write

$$\alpha = 1 + \delta,$$

where δ is a positive number. Using the binomial theorem, we write

$$\alpha^n = (1 + \delta)^n = 1 + \cdots + \frac{n(n-1)\cdots(n-k)}{(k+1)!}\delta^{k+1} + \cdots.$$

The terms omitted on the right side are all positive. Consider in particular the coefficient

$$a = \frac{n(n-1)\cdots(n-k)}{(k+1)!}.$$

Its numerator contains $k+1$ factors. If we suppose that $n > 2k$, we see that each of these factors is larger than $\frac{1}{2}n$, and it follows that

$$\frac{\alpha^n}{n^k} > \frac{1}{n^k}(\tfrac{1}{2}n)^{k+1}\frac{\delta^{k+1}}{(k+1)!} = \frac{n}{2^{k+1}(k+1)!}\delta^{k+1}.$$

Therefore the quantity $\dfrac{n}{n^k}$ grows without bound as n increases, and so (15) is proved. As already remarked, (16) holds as well.

We can improve considerably on (16). Not only is the quantity $n^k \alpha^n$ infinitely small, but it goes to zero rapidly, in fact faster than a certain decreasing geometric progression. Let us prove this. Since $\alpha < 1$, there is a number γ such that

$$\alpha < \gamma < 1.$$

Then the quantity $n^k \dfrac{\alpha^n}{n} = n^k \left(\dfrac{\alpha}{\gamma}\right)^n$ is infinitely small, since $\dfrac{\alpha}{\gamma} < 1$. (Use (16).)
Therefore, for $0 < \alpha < 1$, we have

$$n^k \alpha^n = o(\gamma^n), \tag{17}$$

for every number γ such that $\alpha < \gamma < 1$.

We consider another infinitely small quantity that is important in analysis, namely the quantity

$$\frac{\alpha^n}{n!}, \tag{18}$$

where α is an arbitrary positive number. We will prove not only that the quantity (18) goes to zero with increasing n but that its terms are less than those of a certain decreasing geometric progression. We will prove that for every positive number α, there is a number γ such that $0 < \gamma < 1$ for which the relation

$$\frac{\alpha^n}{n!} = O(\gamma^n) \tag{19}$$

obtains. To do this, we first choose the smallest (necessarily nonnegative) integer k for which the inequality

$$\alpha < k+1$$

holds. We now define

$$\gamma = \frac{\alpha}{k+1}.$$

It is clear that $0 < \gamma < 1$. Now consider integers $n \geq k$. For these values of n, we have $n = k + l$, where $l \geq 0$. We may now write

$$\frac{\alpha^n}{n!} = \frac{\alpha^{k+l}}{(k+l)!} = \frac{\alpha^k}{k!} \cdot \frac{\alpha^l}{(k+1)\cdots(k+l)} \leq \frac{\alpha^k}{k!} \cdot \frac{\alpha^l}{(k+1)^l}$$

$$= \frac{\alpha^k}{k!} \gamma^l = \frac{\alpha^k}{k!\,\gamma^k} \cdot \gamma^{k+l} = \frac{\alpha^k}{k!\,\gamma^k} \gamma^n.$$

For $n \geq k$, we therefore have

$$\frac{\alpha^n}{n!} \leq b\,\gamma^n, \quad \text{where } b = \frac{\alpha^k}{k!\,\gamma^k}.$$

For $n < k$, we consider the putative inequality

$$\frac{\alpha^n}{n!} \leq c\,\gamma^n, \tag{20}$$

where c is a positive number to be determined. Since the values of n in (20) run only through the finite set $1, 2, \ldots, k-1$, we can certainly choose a number c which makes (20) valid. Finally define a as the larger of b and c. We obtain the inequality

$$\frac{\alpha^n}{n!} \leq a\,\gamma^n$$

for all positive integers n, and this proves (19).

The relation (19) states that the quantity $n!$ grows faster than the quantity α^n for all positive numbers α. Taking reciprocals, we may say that the quantity $\dfrac{1}{n!}$ goes to zero faster than the quantity α^n for all positive numbers $\alpha < 1$.

Example. Let α be a positive number. The expression α^n, "alpha to the power n", has an immediate meaning for all positive integers n. In algebra we also define α^n for $n = 0$ and for negative integers n, namely $\alpha^0 = 1$ and $\alpha^n = \dfrac{1}{\alpha^{-n}}$ for negative integers n. In much the same way, we define $\alpha^{r/s}$ for rational numbers $\dfrac{r}{s}$ by using positive s^{th} roots. The definition of α^x for an arbitrary real number x is more delicate. Later on, we shall go into this question in some detail. For the moment let us suppose that we can define α^x

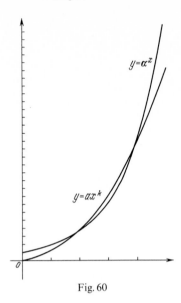

Fig. 60

for all real numbers x, positive negative, or zero, for any fixed positive α:

$$y = \alpha^x. \tag{21}$$

We suppose for present purposes that $\alpha > 1$. Along with the function (21) we consider the function

$$y = a\,x^k, \tag{22}$$

where a is a positive constant and k is a fixed positive integer. We sketch the graphs of both functions for $x \geq 0$ in Fig. 60. For large positive integer values of x, we proved in (15) that the graph of the function (21) will lie above the graph of the function (22). We may suppose that this occurs also for all sufficiently large real values of x. Now consider what happens for $\alpha > 1$ but close to 1, and for large values of the constant a in (22). The curve (22) begins at the origin $(0,0)$, while the curve (21) begins at the point $(0,1)$. In spite of this, the curve (22) climbs faster than the curve (21) for small positive values of x, so that the two curves intersect for some positive x. But then the curve (21) gradually catches up with the curve (22) and so it intersects (22) a second time and then crosses it, remaining thereafter above the curve (21). The phenomenon is illustrated in Fig. 60.

We consider the particular case

$$y = 2^x \tag{23}$$

and

$$y = 4\,x. \tag{24}$$

Exactly this phenomenon occurs for these two curves. The second in-tersection of the two graphs occurs for $x=4$ and $y=16$. It is a nonstandard problem of elementary mathematics to solve the equation

$$2^x = 4x, \tag{25}$$

that is, to find the points of intersection of the graphs (23) and (24). Sometimes students take the point of view that one is to find a formula like the formula for solving quadratic equations that will produce the solution $x=4$ for (25). Such an interpretation is faulty, since inspection shows in-stantly that $x=4$ is a solution of (25). The real problem with (25) is to find a solution less than 4. This solution can be found only by approximation. That there is a solution is evident from looking at the graphs of the functions (23) and (24).[1]

§14. Cauchy's Convergence Criterion

We have already mentioned that curves whose length is not a rational number have been known since ancient times. Recall that rational numbers are numbers of the form $\dfrac{a}{b}$ for integers a and b. We have already cited the example of the length of the hypotenuse of an isosceles right triangle whose sides have length 1. We write the length of this hypotenuse as $\sqrt{2}$. Long after the times of the ancient Greeks, who knew that $\sqrt{2}$ is irrational, it was discovered that the length of a circle of diameter 1 is irrational. This number is nowadays denoted by the symbol π. The set of all irrational numbers is of course infinite. For example, all numbers of the form $a\sqrt{2}$, where a is an integer, are irrational. Mathematicians of the 19th century were faced with the problem of describing, or if you will, constructing, all of the numbers needed for geometry and analysis. In describing these numbers, one must start from somewhere. It is natural to start with the rational numbers, which may be supposed to be fully known and reasonably understandable. One can construct all of the numbers needed in mathematics by various different methods. To give complete proofs of all the important properties of these newly constructed numbers, with any of the methods available for their construction, is a formidable task. In this section I will give one of the constructions of irrational numbers, but will not enter into the details of the proofs of their properties. I will give only sketches of the proofs. A reader who really wants to will be able to reconstruct these proofs completely.

---

[1] *Translator's note.* A few minutes' work with a pocket calculator shows that the solution $x' < 4$ satisfies the inequalities

$$0.30990693 < x' < 0.30990694.$$

However, I do not recommend this, since the proofs strike me as rather boring.

We gave a definition of convergence in §12. It specifies when a given sequence of numbers

$$r_1, r_2, \ldots, r_n, \ldots \tag{1}$$

does or does not converge to a limit r. It is of extreme importance to formulate a condition for convergence in such wise that it answers the question, Does (1) converge to some number?, when we do not know the number to which (1) converges. Cauchy's convergence criterion meets this test.

Cauchy's Criterion. Suppose that for every positive number δ, there exists a positive integer v (depending upon δ) such that for all

$$p > v \quad \text{and} \quad q > v \tag{2}$$

all of the terms r_p and r_q of the sequence (1) satisfy the inequality

$$|r_p - r_q| < \delta. \tag{3}$$

Then the sequence (which may be real or complex) is said *to satisfy Cauchy's criterion*. One also says simply that (1) is a *Cauchy sequence*.

Let us prove that Cauchy's criterion is a necessary condition for convergence. Suppose that the sequence (1) converges to the limit r. Define the positive number ε as $\frac{1}{2}\delta$. The definition of convergence (§12, in particular (11)) shows that there is a positive integer v such that for $n > v$, we have

$$|r_n - r| < \varepsilon.$$

Hence if p and q are integers satisfying (2), we have the two inequalities

$$|r_p - r| < \varepsilon \quad \text{and} \quad |r_q - r| < \varepsilon.$$

From this we find that

$$|r_p - r_q| = |r_p - r + r - r_q| \leqq |r_p - r| + |r - r_q| < 2\varepsilon = \delta.$$

That is, if (1) converges to a limit, it is a Cauchy sequence. We did this for the real and complex cases simultaneously: there is no difference in the proofs. This proof of necessity is very simple.

The proof of sufficiency is a wholly different matter. We do not so much encounter difficulty in a proof but rather the completely fundamental question of what a number is. We need concern ourselves only with real numbers, since once they are known, we define complex numbers z by the familiar identity $z = x + iy$, where x and y are real.

The concept of a rational number evolved step by step over the centuries. Nevertheless, we may now take the position that rational numbers are completely known. As already noted, the first discovery of irrational numbers came about through the theorem of Pythagoras, and it was a great historical event. The very name "irrational", which is to say not reasonable, shows the alarm that the ancient Greek mathematicians felt at encountering irrational numbers. I will not go into the long and complicated development of the concept of irrational numbers, but will only say that Cauchy's criterion as a sufficient condition for convergence has become a vital necessity in mathematics. Thus real numbers must have the property that every Cauchy sequence of real numbers converges to a real number.

Construction of the Real Numbers. We begin with the point of view that rational numbers are completely known. First of all, then, we must infer that every Cauchy sequence of rational numbers converges to some real number. This limit may be rational or irrational. We write down a Cauchy sequence

$$s_1, s_2, \ldots, s_n, \ldots \tag{4}$$

of rational numbers.[1] If this sequence converges to a rational number, we have no problem. However, if there is no rational number to which (4) converges, then we have to admit that there is an irrational number to which (4) converges. Once we admit this, we have in fact arrived at a new concept of what a number is: that a real number is represented by a Cauchy sequence of rational numbers.

Thus by definition every Cauchy sequence of rational numbers yields a real number, which may be rational or irrational. Let s denote the real number represented by the sequence (4). We must regard the terms s_n of (4) as better and better approximations to the real number s. The closeness of this approximation can easily be estimated. Since (4) is a Cauchy sequence, there exists for every positive number δ a natural number v such that any two terms s_p and s_q with indices greater than v differ from each other by less than δ. We must therefore adopt the point of view that every term s_n of (4) with index exceeding v approximates the real number s to within the amount δ. Using the approximants s_n to the real number s, we can define operations on real numbers and answer all questions arising in connection with the definition of s.

Suppose that along with (4) we have another Cauchy sequence

$$s_1', s_2', \ldots, s_n', \ldots \tag{5}$$

---

[1] *Translator's note.* It should be emphasized that the positive numbers δ occurring in the definition of a Cauchy sequence must be rational numbers: other numbers are not yet at our disposal. The reader should keep this in mind while studying the construction of the real numbers.

of rational numbers. We must first answer the question of whether or not (4) and (5) define the same real number. The answer is obvious. We form the sequence

$$s_1 - s_1', s_2 - s_2', \ldots, s_n - s_n', \ldots \tag{6}$$

of rational numbers. The sequences (4) and (5) define the same real number if and only if the sequence (6) converges to 0. In this case the sequences (4) and (5) are said to be *equivalent*.

The real number s defined by (4) is positive if and only if all of the terms of (4), beginning with a certain index, are greater than some fixed positive rational number.

Suppose that we have a Cauchy sequence of rational numbers

$$t_1, t_2, \ldots, t_n, \ldots,$$

which may or may not be different from (4), that defines a real number t. It is clear how we must define operations on the numbers s and t. The sum $s + t$ is defined by the sequence

$$s_1 + t_1, s_2 + t_2, \ldots, s_n + t_n, \ldots,$$

which is again a Cauchy sequence. The difference $s - t$, the product st, and the quotient $\dfrac{s}{t}$ are defined similarly by the Cauchy sequences for s and t.

The quotient $\dfrac{s}{t}$ can be defined, of course, only when the divisor t is nonzero. Since we know when a real number is positive and also how to subtract one real number from another, we can define inequality between two real numbers.

It turns out that under our definition of real numbers, every Cauchy sequence of real numbers converges to a limit. The proof of this fact is uninteresting, and I will not give it here.

Let us define complex numbers z by the familiar formula $z = x + iy$, where x and y are real numbers. It is then easy to prove that a Cauchy sequence of complex numbers converges to a complex number.

Thus we have constructed the real and complex number systems in such a way that a sequence admits a limit if and only if it is a Cauchy sequence.

Example. Let us set down an infinite decimal

$$a . f_1 f_2 \ldots f_n \ldots . \tag{7}$$

To the right of the decimal point stands a sequence of numbers, each of which is an integer between 0 and 9. The number a to the left of the decimal point is an integer, which may be positive, negative, or zero. We do not write it out in decimal form. The decimal expansion $0.f_1 f_2 \ldots f_n \ldots$ appears to be infinite. However, if the f_n's are ultimately all zero, it is a finite decimal fraction. An infinite decimal of the form (7) generates an infinite sequence of finite decimal fractions:

$$
\begin{aligned}
s_0 &= a, \\
s_1 &= a.f_1, \\
s_2 &= a.f_1 f_2, \\
&\ldots \\
s_n &= a.f_1 f_2 \ldots f_n, \\
&\ldots
\end{aligned}
\tag{8}
$$

Here s_n is the finite decimal fraction with integer part a and with exactly n entries $f_1, f_2, \ldots, f_n$ to the right of the decimal point. We have already looked at such expansions (see §12, (8) and (9) and the discussion there). We can now show that the sequence (8) of rational numbers is Cauchy sequence. Let p and q be two natural numbers such that $p < q$. We have the evident inequality

$$
s_q - s_p < \frac{1}{10^p}.
$$

Thus, if both p and q are greater than the natural number v, we have

$$
|s_q - s_p| < \frac{1}{10^{v+1}},
\tag{9}
$$

which implies that the sequence (8) is a Cauchy sequence. By our agreement, therefore, (8) defines a certain real number s, which may be rational or irrational. It is now natural to suppose that a real number s is written in the form of a decimal (7). However, it remains to be shown that this can be done for all real numbers s. We will deal with this matter in §15.

§15. Applications of Cauchy's Convergence Criterion

The sufficiency of Cauchy's convergence criterion enables us in a great variety of situations to prove the existence of or to construct a number that we need. We present in this section a number of examples of this genre. We will also give a precise definition of the length of a circular arc.

Upper and Lower Bounds. Every real number x can be represented as a point on the axis of abscissas, namely, the point whose abscissa is the number x. We denote this point also by the symbol x. The representation of numbers as points on a line gives them a certain geometric concreteness. We will employ both numerical and geometric terminology in the following discussion.

Consider a finite decimal

$$a.f_1 f_2 \cdots f_n, \tag{1}$$

in which the number to the left of the decimal point is an arbitrary integer and to the right of the decimal point stand exactly n numbers. Every such finite decimal (1) can be written in the form $\dfrac{b}{10^n}$, where b is some integer. Conversely, every fraction of the form $\dfrac{b}{10^n}$ can be written in the form of a finite decimal (1). For $n=0$, the number appearing in (1) is an integer. The points corresponding to integers on the line are also called integers or integral. We can easily represent them geometrically. They are said to form *the lattice of integers on the line.* Numbers of the form $\dfrac{b}{10}$ with integral b form a finer lattice, obtained from the lattice of integers by dividing each of its intervals into ten parts of equal length.

In exactly the same way, numbers of the form $\dfrac{b}{10^2}$ with integral b form a lattice finer than the preceding one, again obtained by dividing each interval into ten subintervals of equal length. Thus we obtain a sequence of finer and finer lattices on the line, consisting of numbers of the form $\dfrac{b}{10^n}$ for integers b and $n=0$, 1, 2, ... We number these lattices by the index n, writing Σ_n for the lattice of all numbers $\dfrac{b}{10^n}$ with integral b.

Now let M be any collection, or as one says, *set* of real numbers. We can regard these numbers as points on the line, so that M is also a set of points. Suppose that M is bounded above, which is to say that there is a number c such that the inequality

$$x<c \tag{2}$$

holds for all x in M. Geometrically this means that all of the points of M lie to the left of the point c.

We will now prove that every set M that is bounded above admits a least upper bound or supremum g, which is defined by the following properties. For every positive number ε, there is a point x in M such that

$$x>g-\varepsilon.$$

Next, there is no point x in M for which we have

$$x>g.$$

Such a number g is called *the least upper bound* or *supremum* of the set M. (It is clear that there cannot be two distinct numbers g with the two properties stated.) Geometrically, the supremum g is described as follows. First, if one moves the point g to the left by ever so small an amount ε (that is, one considers the point $g - \varepsilon$), there is at least one point of M lying to the right of $g - \varepsilon$. Also there is no point of the set M that lies to the right of g.

We construct the supremum g of the set M as follows. We will compare geometrically the set M and points of the lattice Σ_n. Let p be a point of the lattice Σ_n lying to the right of the point c (see (2)), so that all of the points of M lie to the left of the point p. On the other hand, there is a point q of the lattice Σ_n for which there are points of M lying to the right of q. Between q and p there are only a finite number of points of the lattice Σ_n. Hence there is a unique point of Σ_n lying farthest to the right such that there are points of M lying to the right of this point. Let this point be denoted by s_n. We define the points s_n for $n = 0$, 1, 2, ..., and so obtain the sequence of numbers

$$s_0, s_1, \ldots, s_n, \ldots . \tag{3}$$

Plainly each s_n coincides with its predecessor or is to the right of its predecessor ($n = 1$, 2, ...). It is also clear that the points of the sequence (3) cannot all ultimately be the same point. This is because s_n is so chosen that there are points of M strictly to the right of s_n. Hence for some index n' greater than n, the point $s_{n'}$ must lie strictly to the right of s_n. Therefore the sequence (3) of finite decimals is a sequence of the form listed in §14, formula (8), and so may be obtained from an infinite decimal of the form §14, formula (7). This decimal expansion is in fact infinite, since the s_n's increase to the right for arbitrarily large values of n. The Cauchy sequence (3) defines a real number g. It is easy to see that g is the supremum of M. Suppose that M consists of a single point, say m. Then the supremum of M is m, and the expansion (3) gives an infinite decimal expansion for the number m. Thus every real number admits an infinite decimal expansion (see the remark at the end the Example in §14). Thus we can regard the set of all real numbers as the set of all infinite decimals.

Suppose next that we have a set M' of real numbers that is bounded below, that is, there is a real number c' such that

$$c' < x \tag{4}$$

for all x in M'. Let us show that a set M' bounded below admits a *greatest lower bound* or *infimum*. This is the (obviously unique) number g' such that for every positive number ε, there is a number x in M' such that $x < g' + \varepsilon$, and such that there are no numbers x in M' such that $x < g'$. To prove that g' exists, let M be the set of all numbers $-x$ as x runs through M': we call this set $-M'$. Since M' is bounded below, M is bounded above and so admits a supremum g. It is clear that the number $g' = -g$ is the infimum of M'.

Upper and Lower Limits. Let M be an infinite set of real numbers that is bounded both below and above, in the sense of the preceding paragraph. A real number p is called the *upper limit of the set* M if for every positive number ε, there are an infinite number of points belonging to M that are greater than $p-\varepsilon$, while there are only a finite number of points belonging to M that are greater than $p+\varepsilon$. (It is easy to see that p is unique if it exists.) Similarly, a number p' is called the *lower limit of the set* M if for every positive number ε, there are an infinite number of points belonging to M that are less than $p+\varepsilon$, while there are only a finite number of points belonging to M that are less than $p-\varepsilon$. Let us prove that M admits an upper limit. We proceed just as in the construction of the supremum of a set bounded above. We fix the positive integer n and consider all rational numbers of the form $\dfrac{b}{10^n}$ with integral b. Among these numbers, let s_n be the largest such that there are an infinite number of points in M that exceed s_n. It is clear that the sequence

$$s_0, s_1, \ldots, s_n, \ldots$$

is a nondecreasing sequence of finite decimals of the type described in §14, (8). Therefore this sequence is a Cauchy sequence. The limit of this sequence is obviously the upper limit p of the set M. To prove the existence of the lower limit p', take the upper limit of the set $M' = -M$. Plainly p and p' are limit points of the set M, as the term is defined in §12, Example 3. It is also clear that p is the largest of all limit points of M and that p' is the smallest of all limit points of M. It can happens that $p = p'$, and then M of course admits only one limit point, namely p. This occurs, for example, when M is the set of all numbers $\pm\dfrac{1}{n}$, as n runs through the set of all positive integers.

Bounded Sequences and Bounded Sets of Numbers. A sequence

$$r_1, r_2, \ldots, r_n, \ldots \tag{5}$$

of real or complex numbers is said to be *bounded* if there exists a positive number c such that the inequality

$$|r_n| < c$$

holds for all of the indices n. Similarly, a set M of real or complex numbers is said to be *bounded* if there exists a positive number c such that the inequality

$$|x| < c$$

holds for all numbers x in M. We will now prove that every bounded sequence (5) of numbers admits a convergent subsequence. We will then prove that every infinite bounded set M admits a limit point (proved in the preceding paragraph for sets M of real numbers).

We have already remarked that a given number may appear in the sequence (5) with different indices. We deal first with the cases in which the sequence (5) takes on only a finite number of distinct values. In this case, there is a number r that appears as the term r_n for an infinite number of distinct indices n. Write all of the terms of the sequence (5) that are equal to r in the order in which they occur. We obtain a subsequence

$$r_{n_1}, r_{n_2}, \ldots, r_{n_k}, \ldots \tag{6}$$

of (5) all the terms of which are equal to r. The subsequence (6) trivially converges to r, and so our assertion holds in this case.

We now suppose that the values assumed by the sequence (5) are an infinite set M. Consider first the case in which (5) is a sequence of real numbers, i.e., M is a set of real numbers. The set M is bounded (this is what is meant by saying that (5) is bounded) and infinite and by the preceding paragraph, M admits at least one limit point p. It was proved in §12, Exmaple 3, that one can choose a sequence

$$\sigma_1, \sigma_2, \ldots, \sigma_n, \ldots \tag{7}$$

of distinct points of M that converges to the number p. We wish to rearrange the order of the terms in (7) so that (7) becomes a subsequence of (5) and still converges to p. To do this, we proceed as follows. Look at the number r_1. If it does not appear in the sequence (7), we throw it out. If it does occur somewhere in (7), we write it as r_{n_1}. Next look at the number r_2. If it occurs nowhere in the sequence (7), we throw it out. If it does occur somewhere in (7), we write it as r_{n_2}. We examine in the same way every term r_n of (5), each in its order. We thus obtain an infinite subsequence

$$r_{n_1}, r_{n_2}, \ldots, r_{n_k}, \ldots \tag{8}$$

of (5). Our procedure for choosing the numbers r_{n_k} never falters, because the sequence (7) has an infinite number of different entries (in fact, no two of them are the same). At the same time, the values assumed by (8) are exactly the values assumed by (7). Every ε-neighborhood of the point p contains all of the numbers σ_n having sufficiently high index n. Let us go out in the sequence (8) until we have eliminated all values of (7) with indices l such that σ_l does not belong to the given ε-neighborhood of p. From this point on in (8), all of the r_{n_k} lie in the given ε-neighborhood of p, and so the sequence (8) converges to p.

Now consider the case in which the sequence (5) admits an infinite number of different values, but these values are not necessarily real. We write

$$r_n = s_n + i t_n$$

for all n and so obtain two real sequences

$$s_1, s_2, \ldots, s_n, \ldots \tag{9}$$

and

$$t_1, t_2, \ldots, t_n, \ldots \tag{10}$$

At least one of the sequences (9) and (10) must have an infinite number of distinct values (otherwise the complex numbers r_n would assume only a finite number of distinct values). With no loss of generality, we suppose that the sequence (9) assumes an infinite number of distinct values. Since it is a real sequence, we use what we have just proved to assert that the sequence (9) admits a subsequence

$$s_{n_1}, s_{n_2}, \ldots, s_{n_k}, \ldots \tag{11}$$

with all values different and converging to a real number s. Now look at the sequence

$$t_{n_1}, t_{n_2}, \ldots, t_{n_k}, \ldots$$

of the sequence (10). It is bounded and so by our previous discussion it admits a subsequence

$$t_{m_1}, t_{m_2}, \ldots, t_{m_k}, \ldots$$

that converges to a number t. Here the positive integers

$$m_1, m_2, \ldots, m_k, \ldots$$

are strictly increasing with k and are a subsubsequence of the subsequence

$$n_1, n_2, \ldots, n_k, \ldots$$

The sequence

$$s_{m_1}, s_{m_2}, \ldots, s_{m_k}, \ldots$$

is therefore a subsequence of the sequence (11) and consists of all different entries. Finally we write

$$r_{m_k} = s_{m_k} + i t_{m_k}.$$

The sequence

$$r_{m_1}, r_{m_2}, \ldots, r_{m_k}, \ldots \tag{12}$$

has all entries different and converges to the complex number $r = s + it$. At the same time, (12) is a subsequence of the original sequence (5).

We now look at a bounded infinite set M of complex numbers. We wish to prove that it admits a limit point. We select any sequence with values taken from M, say

$$r_1, r_2, \ldots, r_n, \ldots \tag{13}$$

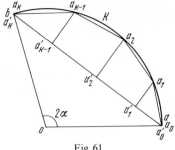

Fig. 61

all of whose values are different. From (13) we can choose a subsequence with all values different that converges to some complex number r. This number is a limit point of the set M.

This completes our proof.

The Length of a Circular Arc. Now that we have explained what a real number is, it is natural to explore what we mean by the length of a curve. We suppose that we know what we mean by the length of a straight line segment. Given a line segment with endpoints p and q (in the plane) let $l(p, q)$ denote the length of this line segment. We consider here only circular arcs. The case of a general curve differs only superficially from the case of a circular arc. (We will treat it later on, in §30.) Consider a plane P and in it a circle of radius 1 and center at o. We choose an arc ab of the circle so that the angle aob subtended by this arc does not axceed 120° (see Fig. 61). Let 2α be the measure of this angle, so that $\alpha \leqq 60°$. This restriction is inessential. We will also suppose that the motion from a to b along the arc is counterclockwise. We wish to give an exact definition of the length of the arc ab.

On the arc ab, we choose a finite sequence of points

$$a_0, a_1, \ldots, a_k \quad \text{with} \quad a_0 = a \text{ and } a_k = b. \tag{14}$$

We will suppose that the direction from a_{i-1} to a_i along the arc ab is counterclockwise. Thus the points (14) follow one after the other along the arc ab and divide it up into pieces. Let A denote the broken line $a_0 a_1 a_2 \ldots a_k$, and let $l(A)$ denote its length, that is, the sum of the length of the line segments that comprise it. We say that A rests on the chord ab. Comparing the length $l(A)$ with the length $l(a, b)$ of the chord ab, we see at once that

$$l(a b) \leqq l(A).$$

This inequality, however, does not suffice for our purposes. Let

$$a'_0, a'_1, a'_2, \ldots, a'_k$$

be the projections of $a_0, a_1, a_2, \ldots, a_k$ respectively onto the chord ab (along lines perpendicular to the chord ab). It is clear that

$$l(a\,b) = l(a'_0\,a'_1) + l(a'_1\,a'_2) + \cdots + l(a'_{k-1}\,a'_k).$$

Let β_i be the angle between the chord $a_{i-1}a_i$ and a line parallel to the chord ab drawn through the point a_{i-1}. It is then clear that

$$l(a_{i-1}\,a_i) = \frac{l(a'_{i-1}\,a'_i)}{\cos\beta_i} < \frac{l(a'_{i-1}\,a'_i)}{\cos\alpha}.$$

Let us sum these inequalities over i from 1 to k. We obtain

$$l(a\,b) \leq l(A) < \frac{l(a\,b)}{\cos\alpha}. \tag{15}$$

We now write $x = l(a\,b)$ and consider the difference

$$d = \frac{x}{\cos\alpha} - x.$$

Observe that $x = 2\sin\alpha$ and that hence $\cos^2\alpha = 1 - \dfrac{x^2}{4}$. We can now estimate the size of d:

$$d = x\left(\frac{1}{\cos\alpha} - 1\right) < x\left(\frac{1}{\cos^2\alpha} - 1\right) = x\frac{\sin^2\alpha}{\cos^2\alpha} \leq \frac{x\sin^2\alpha}{\frac{1}{4}} = x \cdot x^2.$$

We can now replace (15) by

$$x \leq l(A) < x(1 + x^2). \tag{16}$$

We now divide up each arc $a_{i-1}a_i$ into smaller pieces, just as we did with the arc ab. We will not use new letters to denote the points of subdivision. We denote the broken line obtained this way by B_i and note that it rests on the chord $a_{i-1}a_i$. Let $l(B_i)$ denote the length of the broken line B_i. Taking all of the broken lines $B_1, B_2, \ldots, B_k$ together, we obtain a broken line B, whose length we denote by $l(B)$, and we say that the broken line B rests on the broken line A. We now write

$$x_i = l(a_{i-1}\,a_i).$$

Applying the inequalities (16), which hold for any chord, to the chord $a_{i-1}a_i$, we find the inequalities

$$x_i \leq l(B_i) < x_i(1 + x_i^2) \leq x_i(1 + (\delta(A))^2), \tag{17}$$

where $\delta(A)$ is the largest of the numbers x_1, x_2, ..., x_k, that is, the maximum of the lengths of the line segments that comprise the broken line A. Summing the inequalities (17) for all i from 1 to k, we find

$$l(A) \leq l(B) < l(A)(1 + (\delta(A))^2). \tag{18}$$

We also note that all of the broken lines resting on the original chord ab have bounded lengths. In fact, from the second inequality in (15), we find that

$$l(A) < \frac{l(ab)}{\cos \alpha} = c,$$

where the number c depends only on the chord ab. From (18) we now infer that

$$|l(B) - l(A)| < c(\delta(A))^2, \tag{19}$$

where A is any broken line resting on the chord ab and B is any broken line resting on the broken line A.

Suppose now that A and A' are two broken lines that rest on the chord ab. Consider all of the vertices appearing either in A or in A' and list them in the order that they appear as one goes along the arc ab in the counterclockwise direction from a to b. We get the vertices for a broken line B that rests upon both A and A'. Along with (19) we therefore have

$$|l(B) - l(A')| < c(\delta(A'))^2. \tag{20}$$

Combining (19) and (20), we find that

$$|l(A) - l(A')| < c[(\delta(A))^2 + (\delta(A'))^2]. \tag{21}$$

We now construct a sequence of broken lines

$$A_0 = ab, A_1, ..., A_n, ...$$

where the first broken line is simply the chord ab and all of the following broken lines rest upon their predecessors. We suppose as well that

$$\lim_{n \to \infty} \delta(A_n) = 0. \tag{22}$$

From the inequality (19) we see that the sequence of positive numbers

$$l(A_0), l(A_1), ..., l(A_n), ... \tag{23}$$

is a Cauchy sequence. We denote its limit by $2s$:

$$\lim_{n \to \infty} l(A_n) = 2s.$$

It is easy to see that this limit $2s$ depends only on the arc ab, and not upon our choice of the sequence $A_0, A_1, \ldots, A_n, \ldots$, which we happened to choose. Suppose in fact that we choose any other sequence

$$A_0' = ab, A_1', \ldots, A_n', \ldots.$$

In view of (21), we have

$$|l(A_n) - l(A_n')| < c[(\delta(A_n))^2 + (\delta(A_n'))^2].$$

The right side of this inequality goes to 0 as n goes to ∞, and so both the sequence (23) and the sequence

$$l(A_0'), l(A_1'), \ldots, l(A_n'), \ldots$$

have the same limit, $2s$. We define this limit $2s$ as the length of the arc ab. We also define the size of the angle aob as the arc length $2s$. This gives the radian measure of the angle aob.

From (15) we have

$$x = 2\sin s \leq l(A_n) < \frac{x}{\cos s} = \frac{2\sin s}{\cos s}.$$

Going to the limit as $n \to \infty$ and dividing the resulting inequality by 2, we find that

$$\sin s \leq s < \tan s. \tag{24}$$

Suppose now that $s_1, s_2, \ldots, s_n, \ldots$ is a sequence of positive numbers with limit 0. The inequalities (24) show that

$$\lim_{n \to \infty} \frac{\sin s_n}{s_n} = 1 \quad \text{and} \quad \lim_{n \to \infty} \frac{\tan s_n}{s_n} = 1.$$

Writing t_n for the number $\tan s_n$, we obtain

$$\lim_{n \to \infty} \frac{\arctan t_n}{t_n} = 1. \tag{25}$$

Here we may suppose that $t_1, t_2, \ldots, t_n, \ldots$ is an arbitrary sequence of positive numbers with limit 0.

The length of a circle of radius 1 is therefore a well defined number, which is written as 2π. The number π has been calculated to a huge number of decimal places. Here we note only that $3.1415926 < \pi < 3.1415927$.

Example 1. Let

$$s_1, s_2, \ldots, s_n, \ldots \tag{26}$$

be a nondecreasing bounded sequence of real numbers, that is, $s_n \leq s_{n+1}$ and $s_n < c$ for $n = 1, 2, \ldots$, where c is a constant. Let g be the supremum of the set of all numbers (26). It is easy to see that

$$\lim_{n \to \infty} s_n = g.$$

In just the same way, we see that for a nonincreasing sequence of real numbers

$$t_1, t_2, \ldots, t_n, \ldots \tag{27}$$

that is bounded below, we have

$$\lim_{n \to \infty} t_n = g',$$

where g' is the infimum of all of the numbers (27).

§16. Convergent Series

Convergent series play an even greater rôle in mathematics than convergent sequences. The summands in a series may be either numbers or functions. In the latter case, series offer the possibility of beginning with simple and well-known functions, for example, powers of the independent variable, and obtaining much more complicated functions. We will use series first and foremost with this end in mind: see §18. But first we will study series whose summands are numbers, either real or complex.

We write a sum

$$z_1 + z_2 + \cdots + z_n + \cdots \tag{1}$$

with an infinite number of summands. We call this formal sum *a series*. The individual summands $z_n (n = 1, 2, \ldots)$ are numbers, either real or complex. In the event that some assertion about the series (1) has a meaning only for real summands, we will mention this specifically.

We must first explain what we mean by the sum of the series (1). For this we write what are called its *partial sums*

$$s_n = z_1 + z_2 + \cdots + z_n, \tag{2}$$

for all positive integers n. If the sequence

$$s_1, s_2, \ldots, s_n, \ldots \tag{3}$$

of partial sums converges to a limit s (as defined in §1), then we say that the series (1) *converges to the sum s.*

Given a sequence of numbers (3), we can form a series (1) for which (3) is the sequence of partial sums:

$$z_1 = s_1, \; z_2 = s_2 - s_1, \; \ldots, \; z_n = s_n - s_{n-1}, \; \ldots.$$

Thus at first glance it would seem that the theory of series is exactly the same as the theory of convergent sequences. In a sense this is true, but nonetheless there are important differences. First of all, it may happen that a series will arise naturally in solving a certain problem. Then the series must be regarded as a primordial object. On the other hand, if a sequence arises in solving a problem, then it has to be regarded as the primordial object. There are cases in which a sequence materializes in the course of solving a problem, while the question of its convergence can be answered only by studying the corresponding series. There is, however, one important question that makes sense for series but not for sequences. It is well known that the sum of a finite number of summands does not depend upon the order of its terms. This is false for series. It can occur that when we permute the order of summation in (1), we obtain a series with sum different from the sum of the original series. We will show this in Example 1.

Tests for convergence, or as one says, *criteria for convergence*, play an important rôle in the theory of series. The basic necessary and sufficient condition for convergence, due to Cauchy, is a simple reformulation in terms of series of Cauchy's criterion for convergence of a sequence, which we studied in §14. There are also a great many important sufficient conditions for convergence that are not at the same time necessary conditions.

Cauchy's Criterion for Convergence. Consider a series (1). Suppose that for every positive number δ there exists a natural number ν such that the inequality

$$|z_{p+1} + z_{p+2} + \cdots + z_q| < \delta \tag{4}$$

holds for all positive integers p and q such that

$$\nu < p < q.$$

We then say that the series (1) satisfies *Cauchy's criterion for convergence*. This criterion is necessary and sufficient for the series (1) to converge. To see this, note that

$$s_p - s_q = z_{p+1} + z_{p+2} + \cdots + z_q. \tag{5}$$

Thus Cauchy's criterion for the series (1) is equivalent to Cauchy's criterion for convergence of the sequence (3). As we proved in §14, this criterion is both necessary and sufficient for convergence of the sequence (3).

In the particular case where $p = n - 1$ and $q = n$, we have $s_q - s_p = z_n$, and (4) and (5) become

$$|z_n| < \delta,$$

and so, if the series (1) converges, we have

$$\lim_{n \to \infty} z_n = 0. \tag{6}$$

In words, the summands of a convergent series have limit 0.

We draw another inference from Cauchy's convergence criterion. If (1) converges, then the series

$$z_k + z_{k+1} + \cdots + z_{k+n} + \cdots \tag{7}$$

also converges, k being an arbitrary positive integer. Furthermore, for every positive number δ, there exists a natural number v such that the sum of the series (7) has absolute value less than δ for all k that exceed v.

Series with Alternating Sign. There is a useful sufficient condition for convergence of certain series

$$z_1 + z_2 + \cdots + z_n + \cdots \tag{8}$$

with real summands. We suppose that all of the summands with odd indices are positive and that all of the summands with even indices are negative. We suppose in addition that the moduli of the summands decrease strictly:

$$|z_{n+1}| < |z_n|$$

for $n = 1, 2, \ldots$. Finally we suppose that

$$\lim_{n \to \infty} |z_n| = 0.$$

Such a series is said to be of *alternating sign*. All such series converge.

To prove this, we consider first partial sums s_{2k} with even indices. We have

$$s_{2k} = (z_1 + z_2) + (z_3 + z_4) + \cdots + (z_{2k-1} + z_{2k}).$$

Each term on the right side within parentheses is positive, and so the sequence $s_2, s_4, \ldots, s_{2k}, \ldots$ is strictly increasing. We can also write

$$s_{2k} = z_1 + (z_2 + z_3) + (z_4 + z_5) + \cdots + (z_{2k} + z_{2k+1}) - z_{2k+1}.$$

All of the terms within parentheses on the right side of this identity are negative, and the last term $-z_{2k+1}$ is also negative. It follows that

$$s_{2k} < z_1.$$

That is, the sequence of positive numbers

$$s_2, s_4, \ldots, s_{2k}, \ldots$$

is increasing and bounded. As was proved in §4, Example 1, it therefore converges to a limit s. The partial sum s_{2k+1} differs from the partial sum s_{2k} only by the number z_{2k+1}, which has limit zero by our hypothesis. Therefore the sequence of all partial sums converges to the number s.

Thus we have proved that every series with alternating sign converges. It may occur, however, that the sequence of numbers

$$t_k = z_1 + z_3 + z_5 + \cdots + z_{2k-1}$$

increases without bound as k increases. In this case, the order of the terms in the series (8) can be rearranged in such wise that the resulting series converges to an arbitrary preassigned number. This strange fact, that for some convergent series the terms can be rearranged so as to change the sum, leads to the suspicion that series of this kind may display other pathological behavior that would render them of little value for applications. Hence one singles out a class of series, the so-called *absolutely convergent series*, for which all rearrangements yield one and the same sum. We will study these series in §17.

Example 1. Consider a (convergent) series

$$z_1 + z_2 + \cdots + z_n + \cdots \tag{9}$$

with alternating sign. We write two sums:

$$t_k = z_1 + z_3 + \cdots + z_{2k-1}$$
and
$$t_k' = z_2 + z_4 + \cdots + z_{2k}.$$

The first of these sums is positive and increases with k, while the second is negative; its modulus increases with k. Suppose that $t_k \to \infty$ as $k \to \infty$. We must then have $|t_k'| \to \infty$ as $k \to \infty$. To see this, note that

$$s_{2k} = t_k + t_k'.$$

The number s_{2k} converges to a limit s as $k \to \infty$. Plainly, if the numbers t_k get arbitrarily large, the negative numbers t_k' must decrease without bound, converging (as we say) to $-\infty$.

We now consider a series (9) for which $t_k \to \infty$ as $k \to \infty$. We choose any real number t, and will construct a rearrangement of the series (9) that converges to t. Let p_1 be the smallest of the positive integers p for which we have

$$t_p > t.$$

Next let q_1 be the smallest of the positive integers q for which we have

$$t_{p_1} + t_q' < t.$$

Beginning with these initial values of the numbers p_1 and q_1, we define by induction two increasing sequences of natural numbers

$$p_1, p_2, p_n, \dots \qquad (10)$$

and

$$q_1, q_2, \dots, q_n, \dots . \qquad (11)$$

Suppose that we have already defined the numbers p_n and q_n. Then p_{n+1} is the smallest of the positive integers p for which the inequality

$$t_p + t'_{q_n} > t$$

obtains. The number q_{n+1} is the smallest of the positive integers q for which the inequality

$$t_{p_{n+1}} + t'_q < t$$

obtains. Since (9) is a series of alternating sign, our definition requires that its terms go to 0 as $n \to \infty$. From this we infer easily that

$$\lim_{n \to \infty} (t_{p_n} + t'_{q_n}) = t.$$

The sum $t_{p_n} + t'_{q_n}$ can be written in the form

$$t_{p_1} + t'_{q_1} + (t_{p_2} - t_{p_1}) + (t'_{q_2} - t'_{q_1}) + \cdots + (t_{p_n} - t_{p_{n-1}})$$
$$+ (t'_{q_n} - t'_{q_{n-1}}). \qquad (12)$$

Each of the summands in (12) is the sum of certain terms of the series (9). No summand of (9) is repeated in (12), and every summand in (9) occurs in (12) for some value of n. That is, we may replace in (12) each summand by the corresponding sum of terms of the series (9) and obtain a series that contains every summand of the series (9) but arranged in a (possibly) different order. This new series, by its very construction, has sum t, and it is obtained from (9) simply by rearranging the order of summation.

Example 2. We present a specific example of a series of the form considered in Example 1. This is the series

$$1 - \tfrac{1}{2} + \tfrac{1}{3} - \cdots + \frac{(-1)^{n-1}}{n} + \cdots . \qquad (13)$$

The series (13) plainly is of alternating sign, and so converges. The numbers t_k and t'_k are either both bounded or both unbounded ($k = 1, 2, \dots$). Therefore we can prove that t_k is unbounded by proving that

$$\lim_{k \to \infty} (t_k + |t'_k|) = \infty.$$

That is, the series (13) is as in Example 1 provided that the series

$$1 + \tfrac{1}{2} + \tfrac{1}{3} + \cdots + \frac{1}{n} + \cdots \tag{14}$$

diverges, which is to say that its partial sums increase without limit. (We say that its sum is $+\infty$.) Consider the block of terms of the series (14) consisting of the terms with indices n for which

$$2^{l-1} < n \leqq 2^l,$$

where l is a natural number. There are 2^{l-1} terms in this block, and the last term, which is the least of them all, is the number $\dfrac{1}{2^l}$. Therefore the sum of the terms in this block exceeds the number

$$\frac{2^{l-1}}{2^l} = \frac{1}{2}.$$

The whole series (14), beginning with the second summand, can be broken up into disjoint blocks of the form just studied, as l runs from 1 to ∞. Therefore the sum of the series (14) is ∞.

Therefore the numbers t_n for the series (13) are unbounded. From Example 1, we infer that the terms of (13) can be rearranged so that the sum of the new series is any number t that we may choose.

Example 3. Suppose that

$$z_1 + z_2 + \cdots + z_n + \cdots \tag{15}$$

and

$$w_1 + w_2 + \cdots + w_n + \cdots \tag{16}$$

are two convergent series, with sums s and t respectively. Then the series obtained by termwise addition, that is,

$$(z_1 + w_1) + (z_2 + w_2) + \cdots + (z_n + w_n) + \cdots, \tag{17}$$

is also convergent and has $s + t$ as its sum.

To prove this, write the partial sums

$$s_n = z_1 + \cdots + z_n$$

and

$$t_n = w_1 + \cdots + w_n.$$

Convergence of (15) and (16) means that for every positive number δ, there exists a natural number v such that

$$|s_n - s| < \tfrac{1}{2}\delta \quad \text{and} \quad |t_n - t| < \tfrac{1}{2}\delta$$

for all $n > v$. Plainly the nth partial sum of the series (17) is $s_n + t_n$. For all $n > v$, we have

$$|(s_n + t_n) - (s + t)| < \delta,$$

and therefore the series (17) converges to $s + t$.

§17. Absolutely Convergent Series

A series

$$z_1 + z_2 + \cdots + z_n + \cdots \tag{1}$$

is said to be *absolutely convergent* or to *converge absolutely* if the series of moduli of its summands

$$|z_1| + |z_2| + \cdots + |z_n| + \cdots \tag{2}$$

converges.

We must first prove that an absolutely convergent series converges. We will use Cauchy's convergence criterion. For every positive number δ, there exists a natural number v such that the inequality

$$|z_{p+1}| + |z_{p+2}| + \cdots + |z_q| < \delta$$

obtains for all integers p and q such that $v < p < q$. Since

$$|z_{p+1} + z_{p+2} + \cdots + z_q| \leqq |z_{p+1}| + |z_{p+2}| + \cdots + |z_q| < \delta,$$

we see that Cauchy's convergence criterion holds for the series (1) if it holds for the series (2). That is, (1) converges.

The foregoing simple argument is closely connected with a property enjoyed by absolutely convergent series that fails for convergent but non-absolutely convergent series. Let δ be an arbitrary positive number. There exists a natural number v such that for all finite sequences of integers such that $v < p_1 < p_2 < \cdots < p_k$, the inequality

$$|z_{p_1} + z_{p_2} + \cdots + z_{p_k}| < \delta \tag{3}$$

obtains.

To prove this, we need only write the inequalities

$$|z_{p_1}+z_{p_2}+\cdots+z_{p_k}|\leqq|z_{p_1}|+|z_{p_2}|+\cdots+|z_{p_k}|$$
$$\leqq|z_{p+1}|+|z_{p+2}|+\cdots+|z_q|<\delta,$$

where $p+1=p_1$ and $q=p_k$.

Independence of the Order of the Summands. We will now prove that the sum s of an absolutely convergent series does not depend upon the order of its summands. To do this, we write the series

$$z_1'+z_2'+\cdots+z_n'+\cdots, \tag{4}$$

obtained from the series (1) by some rearrangement of its summands. We write the partial sums for both series:

$$s_n=z_1+z_2+\cdots+z_n \quad \text{and} \quad s_n'=z_1'+z_2'+\cdots+z_n'.$$

Now let δ be an arbitrary positive number. Choose a natural number v so large that the inequality

$$|s_n-s|<\delta \tag{5}$$

holds for all $n>v$ and also (3) holds for this δ and the chosen v. We now fix $n=v+1$. The partial sum s_n is thus a determined number, and the terms of (1) that occur in s_n are also fixed. Since (4) is a rearrangement of (1), there is a positive integer v' such that the partial sum $s_{v'}'$ contains all of the summands of (1) that occur in s_n. For $n'>v'$, the terms of $s_{n'}'$ that are not terms of s_n are numbers z_q with $q>n$. Therefore (3) shows that

$$|s_{n'}'-s_n|<\delta. \tag{6}$$

The inequalities (5) and (6) together show that

$$|s_{n'}'-s|<2\delta$$

for all $n'>v'$. Note that v' depends upon δ. Now, given an arbitrary positive number ε, define δ as $\frac{1}{2}\varepsilon$. Thus, starting with an arbitrary positive number ε, we find a positive integer v' such that

$$|s_{n'}'-s|<\varepsilon$$

for all $n'>v'$. That is, the series (4) and the series (1) have the same sum. This completes our proof.

Multiplication of Series. To multiply two finite sums, we must multiply each summand of the first sum by each summand of the second, and add together all of these products. The same rule applies for multiplying two absolutely convergent series. We write two absolutely convergent series:

$$z_0 + z_1 + \cdots + z_n + \cdots \tag{7}$$

and

$$w_0 + w_1 + \cdots + w_n + \cdots. \tag{8}$$

The first summands are written with indices 0 instead of 1. This is not important, and it makes our later notation more convenient than it would otherwise be. We write the partial sums of these series:

$$s_n = z_0 + z_1 + \cdots + z_n \quad \text{and} \quad t_n = w_0 + w_1 + \cdots + w_n.$$

We write the partial sums for the corresponding series of absolute values as

$$\hat{s}_n = |z_0| + |z_1| + \cdots + |z_n|$$

and

$$\hat{t}_n = |w_0| + |w_1| + \cdots + |w_n|.$$

We also write

$$\lim_{n \to \infty} s_n = s \quad \text{and} \quad \lim_{n \to \infty} t_n = t.$$

We now form all of the products $z_p w_q$ of the summands of our series (7) and (8). We will arrange these products in some way as a sequence to be the summands of an infinite series. One possibility is to write

$$z_0 w_0 + (z_1 w_0 + z_0 w_1) + (z_2 w_0 + z_1 w_1 + z_0 w_2) + \cdots$$
$$+ (z_n w_0 + z_{n-1} w_1 + \cdots + z_0 w_n) + \cdots.$$

The rule for writing the numbers $z_p w_q$ by this system is clear. For every nonnegative integer k, we write in one block the products $z_p w_q$ for which $p + q = k$, in the order $z_k w_0, z_{k-1} w_1, \ldots, z_0 w_k$. These are ordered by $k = 0$, $k = 1, \ldots$.

Thus we can write all of the products $z_p w_q$ as a single sequence. Doing this in any way we like, let us write these products as $g_0, g_1, \ldots, g_m, \ldots$ and form the series

$$g_0 + g_1 + \cdots + g_m + \cdots. \tag{9}$$

Let h_m be the mth partial sum of the series (9):

$$h_m = g_0 + g_1 + \cdots + g_m.$$

We will prove that the series (9) converges absolutely and hence converges with sum h, which is to say

$$\lim_{m \to \infty} h_m = h.$$

We will also prove that

$$h = st.$$

To carry out our proof, we first consider the partial sums of the series of absolute values of (9):

$$\hat{h}_m = |g_0| + |g_1| + \cdots + |g_m|.$$

Each partial sum $\hat{h}_m$ contains only a finite number of summands $|z_p w_q|$, and therefore, for a given m, there is a positive integer n so large that all of the summands of $\hat{h}_m$ appear in the product $\hat{s}_n \hat{t}_n$. From this we see that

$$\hat{h}_m \leq \hat{s}_n \hat{t}_n.$$

Since both $\hat{s}_n$ and $\hat{t}_n$ remain bounded over all n, the partial sums $\hat{h}_m$ are also bounded over all indices m. Therefore the series (9) converges absolutely.

Now consider the partial sums h_m of the series (9). For a given m, take n so large that all of the summands in h_m occur in the product $s_n t_n$. Every number $z_p w_q$ occurring in $s_n t_n$ that does not occur in h_m must have the form g_l for some l, and our choice of n ensures that $l > m$. That is, we have

$$s_n t_n - h_m = g_{p_1} + g_{p_2} + \cdots + g_{p_k},$$

where the indices p_j satisfy the inequalities $m < p_1 < p_2 < \cdots < p_k$. We have already proved that given a positive number δ, there is a natural number v such that the inequality

$$|g_{p_1} + g_{p_2} + \cdots + g_{p_k}| < \delta$$

holds for all indices p_j satisfying $v < p_1 < p_2 < \cdots < p_k$. We can also choose this natural number v so that

$$|s_n - s| < \delta \quad \text{and} \quad |t_n - t| < \delta$$

for all $n > v$. Combining all of this, we find that

$$\lim_{m \to \infty} h_m = \lim_{n \to \infty} s_n t_n = st.$$

This proves our rule for multiplying two absolutely convergent series.

A Basic Formula of Elementary Algebra. We pause to prove an important formula of elementary algebra that we shall need many times in our later work, and in particular in finding the sum of a geometric progression. For two variables z and w and a positive integer k, define

$$f_k(z, w) = z^k + z^{k-1} w + z^{k-2} w^2 + \cdots + z w^{k-1} + w^k. \tag{10}$$

That is, $f_k(z, w)$ is the sum of all of the monomials $z^h w^{k-h}$ as h runs from 0 to k. We wish to prove that

$$z^n - w^n = (z - w) f_{n-1}(z, w) \tag{11}$$

for $n = 2, 3, \ldots$. To do this, let us multiply $f_{n-1}(z, w)$ by z. It is clear from (10) that this product contains all of the monomials $z^k w^{n-k}$ occurring in $f_n(z, w)$ except for w^n. That is, we have

$$f_{n-1}(z, w) z = f_n(z, w) - w^n. \tag{12}$$

In exactly the same way, we have

$$f_{n-1}(z, w) w = f_n(z, w) - z^n. \tag{13}$$

Subtract (13) from (12) to obtain (11).

The Sum of a Geometric Progression. We use (11) to find the sum of a geometric progression

$$s_n = a + aw + aw^2 + \cdots + aw^n, \tag{14}$$

where a and w are complex numbers and $w \neq 1$. It is clear that $s_n = af_n(1, w)$. From (11) we obtain

$$s_n = \frac{a(1 - w^{n+1})}{1 - w}. \tag{15}$$

We use (15) to obtain the sum of the infinite geometric progression

$$a + aw + aw^2 + \cdots + aw^n + \cdots \tag{16}$$

for $|w| < 1$. The nth partial sum of (16) is the finite geometric progression (14). In view of (15), we can take the limit as $n \to \infty$ for the case $|w| < 1$ and obtain

$$\lim_{n \to \infty} s_n = \frac{a}{1 - w}. \tag{17}$$

(See formula (12) in §13.)

The Comparison Test for Absolute Convergence. We present here a very simple though very important sufficient condition for absolute convergence of a series Consider a series

$$z_0 + z_1 + \cdots + z_n + \cdots, \tag{18}$$

the terms of which are either real or complex numbers. We consider another series

$$u_0 + u_1 + \cdots + u_n + \cdots, \tag{19}$$

whose summands are positive real numbers. Suppose that the series (19) converges and that inequalities

$$|z_n| \leq u_n$$

hold for all indices n. Then the series (18) converges absolutely. The proof of this assertion is so simple that there is no need to give it here. We note an important additional fact. If s is the sum of the series (18) and v the sum of (19), then we have

$$|s| \leq v.$$

The geometric progression

$$a + a\alpha + a\alpha^2 + \cdots + a\alpha^n + \cdots \tag{20}$$

is an especially important choice for the series (19): here a is any positive number and α is a number such that $0 < \alpha < 1$. We have already proved that the series converges to the sum v, where

$$v = \frac{a}{1-\alpha}. \tag{21}$$

Thus, if the series (18) satisfies the conditions

$$|z_n| \leq a\alpha^n \quad \text{for } n = 0, 1, 2, \ldots, \tag{22}$$

then the series (18) converges absolutely, and for its sum s we have the inequality

$$|s| \leq \frac{a}{1-\alpha}. \tag{23}$$

We can write all of the inequalities (22) briefly in the form

$$z_n = O(\alpha^n) \tag{24}$$

(see §13). Thus (24) is a sufficient condition for the absolute convergence of the series (18).

Example 1. We will give an equality whose proof is an excellent illustration of the comparison test, which we have just discussed. This equality will also be useful in the following section. Let $c(k)$ be a bounded function of the positive integral variable k, that is, a function satisfying the condition

$$|c(k)| < c,$$

where c is a constant independent of k. Then we have

$$\lim_{k \to \infty} \left(1 + \frac{c(k)}{k^2}\right)^k = 1. \tag{25}$$

To prove this, we first expand the expression after the limit symbol by Newton's binomial theorem:

$$\left(1+\frac{c(k)}{k^2}\right)^k = 1 + z_1 + z_2 + \cdots + z_n + \cdots,$$

where

$$z_n = \frac{k(k-1)\cdots(k-n+1)}{n!}\left(\frac{c(k)}{k}\right)^n = \frac{1}{n!}\cdot\frac{k}{k}\cdot\frac{k-1}{k}\cdots\frac{(k-n+1)}{k}\left(\frac{c(k)}{k}\right)^n.$$

The preceding identities hold for arbitrarily large n, although Newton's binomial formula contains only a finite number of terms. We see at once that

$$|z_n| \leqq \left(\frac{c}{k}\right)^n.$$

In view of (23) we thus have for $k > c$ the estimate

$$|z_1 + z_2 + \cdots + z_n| \leqq \frac{\dfrac{c}{k}}{1-\dfrac{c}{k}} = \frac{c}{k-c}.$$

If follows from this estimate that as $k \to \infty$, the sum $z_1 + z_2 + \cdots + z_n$ goes to zero. This proves (25). We can rewrite (25) in the form

$$\lim_{k\to\infty}\left(1+O\left(\frac{1}{k^2}\right)\right)^k = 1.$$

Example 2. Let us prove that the series

$$1 + \tfrac{1}{4} + \tfrac{1}{9} + \cdots + \frac{1}{n^2} + \cdots \tag{26}$$

converges. To do this, we construct an auxiliary series whose terms exceed those of (26) and whose convergence is obvious. This is the following series:

$$1 + \frac{1}{1\cdot 2} + \frac{1}{2\cdot 3} + \cdots + \frac{1}{(n-1)n} + \cdots. \tag{27}$$

The identities

$$\frac{1}{(n-1)n} = \frac{1}{n-1} - \frac{1}{n}$$

are obvious. Rewriting (27) with them, we obtain the new series

$$1+(1-\tfrac{1}{2})+(\tfrac{1}{2}-\tfrac{1}{3})+\cdots+\left(\frac{1}{n-1}-\frac{1}{n}\right)+\left(\frac{1}{n}-\frac{1}{n+1}\right)+\cdots.$$

The sum of the last series is plainly 2. Since

$$\frac{1}{n^2}<\frac{1}{(n-1)\,n},$$

the comparison test shows that the series (26) converges and that its sum is less than 2.

§ 18. The Function exp(z)

The function exp(z) plays a very important rôle in analysis. It is defined for all complex values by the series

$$\exp(z)=\frac{1}{0!}+\frac{z}{1!}+\frac{z^2}{2!}+\cdots+\frac{z^n}{n!}+\cdots. \tag{1}$$

The series (1) converges absolutely for all complex numbers z, since

$$\frac{|z|^n}{n!}=O(\gamma^n), \tag{2}$$

where $\gamma<1$. This is a sufficient condition for absolute convergence of the series (1), as is shown in § 13, (19), and § 17, (24).

The Fundamental Properties of exp(z). First of all, it is clear that

$$\exp(0)=1. \tag{3}$$

The most important property of the function exp(z) is the following identity:

$$\exp(z+w)=\exp(z)\exp(w). \tag{4}$$

Here z and w are arbitrary complex numbers. To prove (4), we write the series

$$\exp(w)=\frac{1}{0!}+\frac{w}{1!}+\frac{w^2}{2!}+\cdots+\frac{w^n}{n!}+\cdots. \tag{5}$$

Since both of the series (1) and (5) converge absolutely, we can find their product by multiplying them together term by term and adding resulting terms in any order we like. That is, we must set down all of the products

$$\frac{1}{p!\,q!}\,z^p\,w^q,\tag{6}$$

where p and q are arbitrary nonnegative integers, and add them up. We do this by adding all of the products (6) for which $p+q=n$, n being a fixed nonnegative integer. We then multiply the resulting finite sum by $n!$. We obtain all summands of the form

$$\frac{n!}{p!\,q!}\,z^p\,w^q\tag{7}$$

with $p+q=n$ (the numbers p and q are of course nonnegative integers). Newton's binomial theorem (§13, (8)) shows that the sum of all of the terms (7) is

$$(z+w)^n.$$

We get all of the terms (6) as we write these terms with $n=0,1,2$, and so on. Therefore we find that

$$\exp(z)\cdot\exp(w)=\frac{1}{0!}+\frac{z+w}{1!}+\frac{(z+w)^2}{2!}+\cdots+\frac{(z+w)^n}{n!}+\cdots=\exp(z+w).$$

(See the series (1)). Thus we have proved the identity (4).
    From (3) and (4), we see that

$$\exp(-z)=\frac{1}{\exp(z)}\tag{8}$$

for all complex numbers z, since

$$1=\exp(0)=\exp(z+(-z))=\exp(z)\cdot\exp(-z).$$

From this (8) is immediate.
    From (4) and (8) we find that

$$\exp(w-z)=\frac{\exp(w)}{\exp(z)}\tag{9}$$

for all complex numbers w and z. Applying (4) repeatedly, we find that

$$\exp(k\,z)=(\exp(z))^k$$

for all complex numbers z and all positive integers k. From this and (8) we find that

$$\exp(k\,z) = (\exp(z))^k \qquad (10)$$

for all complex z and all integers k.

We will now prove that

$$|\exp(z) - 1| < \frac{\alpha}{1-\alpha} \qquad (11)$$

if $|z| = \alpha < 1$. We have

$$\exp(z) - 1 = \frac{z}{1!} + \frac{z^2}{2!} + \cdots + \frac{z^n}{n!} + \cdots. \qquad (12)$$

For $|z| = \alpha < 1$, it is clear that

$$\frac{|z|^n}{n!} \leq \alpha^n.$$

Therefore the modulus of the series on the right side of (12) does not exceed the value $\dfrac{\alpha}{1-\alpha}$ (see §17, (23)). This proves (11).

From (11) it follows that

$$|\exp(w) - \exp(z)| < |\exp(z)|\,\frac{\alpha}{1-\alpha} \qquad (13)$$

if $|w - z| = \alpha < 1$. In fact, (9) give us

$$\exp(w) - \exp(z) = \exp(z) \cdot (\exp(w - z) - 1),$$

and from this and (11), (13) follows.

From (13) we see that the function $\exp(z)$ is continuous (see Example 1 in §12). Suppose that z_0 is an arbitrary complex number and that $z_1, z_2, \ldots, z_n, \ldots$ is a sequence of complex numbers such that

$$\lim_{n \to \infty} z_n = z_0.$$

For all sufficiently large n, (13) shows that

$$|\exp(z_n) - \exp(z_0)| < |\exp(z_0)|\,\frac{|z_n - z_0|}{1 - |z_n - z_0|}.$$

Take the limit as $n \to \infty$ in this inequality to find that

$$\lim_{n \to \infty} \exp(z_n) = \exp(z_0).$$

Therefore the function $\exp(z)$ is continuous everywhere.

The Function exp(z) for Real Values z = x. We now establish certain properties of the function exp(z) for real values $z = x$ of its argument. First we note that exp(x) is positive for all real values of x. For nonnegative x, all of the terms of the series (1) are real and nonnegative (positive for $x > 0$), and so the sum exp(x) of the series is positive. For negative x, use formula (8) to see that exp(x) is positive. Therefore we have

$$\exp(x) > 0$$

for all real x.

The value of the function exp(x) at 1, exp(1), plays a unique rôle in mathematics. We designate this number by the special symbol e. By formula (1), we have

$$e = \exp(1) = 1 + \frac{1}{1!} + \frac{1}{2!} + \cdots + \frac{1}{n!} + \cdots. \tag{14}$$

Because of its uniquely important place in mathematics, the number e has been computed to an enormous number of places of decimals. Here we remark only that

$$2.7182818 < e < 2.7182819.$$

From (10) and (14) we see that

$$\exp(k) = e^k \tag{15}$$

for all integers k. For all rational numbers r, we also have

$$\exp(r) = e^r. \tag{16}$$

To see this, put $k = l$ and $lz = x$ in formula (10). This gives

$$\exp(x) = \left(\exp\left(\frac{x}{l}\right)\right)^l. \tag{17}$$

Since the numbers exp(x) and $\exp\left(\frac{x}{l}\right)$ are positive, (17) implies that

$$\exp\left(\frac{x}{l}\right) = (\exp(x))^{\frac{1}{l}}.$$

We of course take the positive l^{th} root of the positive number exp(x) on the right side of the last equality. Now set $x = k$ in the last equality and apply (15) to write

$$\exp\left(\frac{k}{l}\right) = (\exp(k))^{\frac{1}{l}} = (e^k)^{\frac{1}{l}} = e^{\frac{k}{l}}.$$

Since every rational number has the form $\frac{k}{l}$, formula (16) is proved.

On the right side of (16) we have the number e to a rational power r. This number is defined by the rules of elementary algebra: one takes powers, the reciprocals of powers, and positive roots. Thus the function $\exp(r)$ for rational values of r has a meaning clear from elementary algebra. We may now try to extend the function e^x, defined for rational values of x, to all real values of x in such a way that it is a continuous function of the real variable x. We can do this by setting

$$e^x = \exp(x). \tag{18}$$

For rational values $x = r$, (18) holds because of (16). The right side of (18) is defined for all real values of x and as we have noted it is a continuous function of x. Therefore (18) provides a continuous extension of e^r to all real values of x.

There is no other continuous extension of e^r over all real values of x. To see this, suppose that x_0 is an arbitrary real number. There is a sequence of rational numbers

$$r_1, r_2, \ldots, r_n, \ldots$$

such that

$$\lim_{n \to \infty} r_n = x_0.$$

Since we require that the function e^x be continuous at the point $x = x_0$, we must have

$$\lim_{n \to \infty} e^{r_n} = e^{x_0}.$$

On the left side of this equality, replace e^{r_n} by $\exp(r_n)$. Since the function $\exp(x)$ is continuous, we have

$$\lim_{n \to \infty} \exp(r_n) = \exp(x_0).$$

It follows that $e^{x_0} = \exp(x_0)$.

We have thus proved the identity

$$e^x = \exp(x) = \frac{1}{0!} + \frac{x}{1!} + \frac{x^2}{2!} + \cdots + \frac{x^n}{n!} + \cdots \tag{19}$$

for all real numbers x. This formula has deep significance. Note that it gives an expression as the sum of a series for numbers e^x that (for rational x at least) are defined purely algebraically, and so permits us to calculate the numbers e^x with great accuracy.

The Function exp(z) for Imaginary Values $z = i\,y$. Having examined the function $\exp(z)$ for real values $z = x$, let us now take it up for pure imaginary values $z = iy$, y being a real number. We will establish the remarkable identity

$$\exp(i\,y) = \cos y + i \sin y, \tag{20}$$

where $\cos y$ and $\sin y$ are the elementary trigonometric functions that are familiar to everyone who has studied trigonometry. We use radian measure to define the size of angles y.

We proceed to prove (20). First of all, we set $z = iy$ in the series (1) that defines $\exp(z)$. The resulting series contains terms of two sorts: those that do not have the imaginary number i as coefficient, and those that do. Since the series (1) converges absolutely, we may collect the terms without i and write them as a separate series, and similarly for the terms with i. We obtain two series with real coefficients:

$$f(y) = 1 - \frac{y^2}{2!} + \cdots + (-1)^m \frac{y^{2m}}{(2m)!} + \cdots \tag{21}$$

and

$$g(y) = y - \frac{y^3}{3!} + \cdots + (-1)^m \frac{y^{2m+1}}{(2m+1)!} + \cdots . \tag{22}$$

Since the series for $\exp(z)$ converges absolutely, so do the series (21) and (22). In view of the results of §16 (see Example 3), we can add them term by term. Consequently we find that

$$\exp(i\,y) = f(y) + i\,g(y).$$

Thus, in proving (20), we will also prove that

$$\cos y = f(y) \quad \text{and} \quad \sin y = g(y). \tag{23}$$

That is, we will obtain power series expansions for $\cos y$ and $\sin y$. We repeat that the series (21) and (22) must converge absolutely, since the series (1) does.

In proving (20), we must find approximations to the functions $f(y)$ and $g(y)$ that hold for small values of y. We will suppose that $|y| < \frac{1}{2}$. The terms of the series (21), beginning with the second, do not exceed in absolute value the terms of the geometric progression

$$|y|^2 + \cdots + |y|^{2m} + \cdots ,$$

and all of the terms of the series (22), beginning with the second, do not exceed in absolute value the terms of the geometric progression

$$|y|^3 + \cdots + |y|^{2m+1} + \cdots .$$

The sum of the first geometric progression is

$$\frac{|y|^2}{1-|y|^2} < \frac{|y|^2}{1-\frac{1}{4}} = \tfrac{4}{3}|y|^2,$$

(see § 17, (23)). The sum of the second geometric progression is

$$\frac{|y|^3}{1-|y|^2} < \frac{|y|^3}{1-\frac{1}{4}} = \tfrac{4}{3}|y|^3.$$

In both of these estimates, we suppose of course that $|y| < \tfrac{1}{2}$. Applying these estimates to the series for $f(y)$ and $g(y)$, we find that

$$|f(y)-1| < \tfrac{4}{3}|y|^2 \quad \text{and} \quad |g(y)-y| < \tfrac{4}{3}|y|^3 \tag{24}$$

for $|y| < \tfrac{1}{2}$.

The idea of our proof of (20) is as follows. In view of (10) we have

$$\exp(i\beta) = \left(\exp\left(\frac{i\beta}{k}\right)\right)^k = \exp(i\alpha))^k,$$

where $\alpha = \dfrac{\beta}{k} = O\left(\dfrac{1}{k}\right)$. We suppose that β is independent of k. Here k is a positive integer: we suppose also that k is even. We approximate the complex number $\exp(i\alpha) = f(\alpha) + ig(\alpha)$ on the basis of the formulas (24). For large enough k, these formulas are applicable, and so we find

$$f(\alpha) = 1 + O\left(\frac{1}{k^2}\right) \quad \text{and} \quad g(\alpha) = \alpha + O\left(\frac{1}{k^3}\right). \tag{25}$$

We raise the number $f(\alpha) + ig(\alpha)$ to the power k and let $k \to \infty$. After a series of calculations we will find that

$$\lim_{k \to \infty} (\exp(i\alpha))^k = \cos\beta + i\sin\beta. \tag{26}$$

This will prove (20) for $y = \beta$.

Let us write the complex number $\exp(i\alpha)$ in trigonometric form:

$$f(\alpha) + ig(\alpha) = \rho(\cos\delta + i\sin\delta),$$

where ρ and δ are functions of the number $\alpha = \dfrac{\beta}{k}$, and hence of the even positive integer k. We also have

$$(\exp(i\alpha))^k = \rho^k[\cos(k\delta) + i\sin(k\delta)],$$

as was noted in §3, p. 21, formula (32). To prove (26), it therefore suffices to prove that

$$\lim_{k \to \infty} \rho^k = 1 \tag{27}$$

and

$$\lim_{k \to \infty} k\,\delta = \beta. \tag{28}$$

Using the approximations (25) and calculating a little, we find

$$\rho^2 = [f(\alpha)]^2 + [g(\alpha)]^2 = \left[1 + O\left(\frac{1}{k^2}\right)\right]^2 + \left[\alpha + O\left(\frac{1}{k^3}\right)\right]^2$$

$$= 1 + O\left(\frac{1}{k^2}\right).$$

Note also that

$$\delta = \arctan \frac{g(\alpha)}{f(\alpha)}.$$

To estimate the quotient $\dfrac{f(\alpha)}{g(\alpha)}$, we consider the difference

$$\frac{g(\alpha)}{f(\alpha)} - \alpha = \frac{\alpha + O\left(\frac{1}{k^3}\right)}{1 + O\left(\frac{1}{k^2}\right)} - \alpha = \frac{\alpha + O\left(\frac{1}{k^3}\right) - \alpha - \alpha \cdot O\left(\frac{1}{k^2}\right)}{1 + O\left(\frac{1}{k^2}\right)}$$

$$= \frac{O\left(\frac{1}{k^3}\right)}{1 + O\left(\frac{1}{k^2}\right)} = O\left(\frac{1}{k^3}\right).$$

Therefore we have

$$\frac{g(\alpha)}{f(\alpha)} = \alpha + O\left(\frac{1}{k^3}\right) = \gamma,$$

where

$$\gamma = O\left(\frac{1}{k}\right).$$

We find accordingly that

$$\delta = \arctan \gamma = \frac{\arctan \gamma}{\gamma} \cdot \left(\alpha + O\left(\frac{1}{k^3}\right)\right). \tag{29}$$

The approximations just obtained for ρ^2 and δ make it easy to prove (27) and (28). We have

$$\lim_{k \to \infty} \rho^k = \lim_{k \to \infty} (\rho^2)^{\frac{1}{2}k} = \lim_{k \to \infty} \left(1 + O\left(\frac{1}{k^2}\right)\right)^{\frac{1}{2}k} = \lim_{l \to \infty} \left(1 + O\left(\frac{1}{l^2}\right)\right)^l,$$

where $l = \frac{1}{2} k$. From (25) in §17, we find that

$$\lim_{l \to \infty} \left(1 + O\left(\frac{1}{l^2}\right)\right)^l = 1.$$

Thus the equality (27) is established. We turn now to (28). By (29), we have

$$\lim_{k \to \infty} k\,\delta = \lim_{k \to \infty} \frac{\arctan \gamma}{\gamma}\left(k\alpha + kO\left(\frac{1}{k^3}\right)\right) = \lim_{k \to \infty} \frac{\arctan \gamma}{\gamma} \lim_{k \to \infty} \left(k\alpha + kO\left(\frac{1}{k^3}\right)\right).$$

In view of formula (25) of §15, we have

$$\lim_{k \to \infty} \frac{\arctan \gamma}{\gamma} = 1,$$

and

$$\lim_{k \to \infty} \left(k\alpha + kO\left(\frac{1}{k^3}\right)\right) = \lim_{k \to \infty} \left(\beta + O\left(\frac{1}{k^2}\right)\right) = \beta.$$

This proves the equality (28). We have therefore proved (26) and therefore also the fundamental formula (20).

Example 1. We can give a different, though equivalent, definition of the function exp(z), already defined by the series (1). In fact, the identity

$$\exp(z) = \lim_{k \to \infty} \left(1 + \frac{z}{k}\right)^k \tag{30}$$

holds for all complex z. Expand the kth power on the right side of (30) by Newton's binomial theorem (pp. 125–126). We obtain

$$\left(1 + \frac{z}{k}\right)^k = 1 + \frac{k}{1!} \cdot \frac{z}{k} + \cdots + \frac{k(k-1)\cdots(k-n+1)}{n!} \cdot \frac{z^n}{k^n} + \cdots. \tag{31}$$

The mth term in the original series (1) is

$$z_m = \frac{z^m}{m!}.$$

The mth term in the series (31) is

$$z_m^* = l_m(k) \frac{z^m}{m!},$$

where

$$l_m(k) = \frac{k(k-1)\cdots(k-m+1)}{m!},$$

so that

$$0 \leqq l_m(k) < 1 \tag{32}$$

and

$$\lim_{k \to \infty} l_m(k) = 1. \tag{33}$$

Accordingly, we may write

$$|(z_{n+1} + z_{n+2} + \cdots) - (z^*_{n+1} + z^*_{n+2} + \cdots)| \leqq |z_{n+1}| + |z_{n+2}| + \cdots$$
$$= \delta_n. \tag{34}$$

Since the series (1) converges absolutely, we see that

$$\lim_{n \to \infty} \delta_n = 0. \tag{35}$$

We also write

$$\begin{aligned} s_n &= z_0 + z_1 + \cdots + z_n, \\ s^*_n &= z^*_0 + z^*_1 + \cdots + z^*_n. \end{aligned} \tag{36}$$

From (33) and (34) we see that

$$|s^*_n - s_n| = \varepsilon_k(n), \tag{37}$$

and for each fixed n we have

$$\lim_{k \to \infty} \varepsilon_k(n) = 0. \tag{38}$$

Combine (34), (36), and (37) to obtain

$$\left| \left(1 + \frac{z}{k}\right)^k - \exp(z) \right| \leqq \delta_n + \varepsilon_k(n). \tag{39}$$

Now let α be an arbitrarily small positive number, and let n be so large that $\delta_n < \frac{1}{2}\alpha$ (recall (35)). Fix this value of n, and having done this, choose k so large that $\varepsilon_k(n) < \frac{1}{2}\alpha$ (recall (38)). From (39) we now obtain

$$\left| \left(1 + \frac{z}{k}\right)^k - \exp(z) \right| < \alpha. \tag{40}$$

That is, for arbitrarily small positive α, one can find k so large that (40) holds. This implies of course the limit formula (30).

The relation (30) is of special interest for $z = 1$. In this case, we have

$$e = \lim_{k \to \infty} \left(1 + \frac{1}{k}\right)^k.$$

In fact, the number e was first defined in this fashion.

§19. The Elementary Transcendental Functions

The following functions are called *the elementary transcendental functions:* the exponential function a^x, where a is a positive number; the trigonometric functions $\sin x$, $\cos x$, $\tan x$, $\text{ctn}\, x$; and the functions inverse to the foregoing. $\log_a x$, $\arcsin x$, $\arccos x$, $\arctan x$, and $\text{arcctn}\, x$.

In the preceding section, we began with the exponential function $\exp(z)$ and found power series expansions for the functions e^x, $\cos x$, and $\sin x$:

$$e^x = 1 + \frac{x}{1!} + \frac{x^2}{2!} + \cdots + \frac{x^n}{n!} + \cdots, \tag{1}$$

$$\cos x = 1 - \frac{x^2}{2!} + \cdots + (-1)^n \frac{x^{2n}}{(2n)!} + \cdots, \tag{2}$$

$$\sin x = x - \frac{x^3}{3!} + \cdots + (-1)^n \frac{x^{2n+1}}{(2n+1)!} + \cdots \tag{3}$$

(see §18, (19) and (21)–(23)). These series converge absolutely for all real values x.

These power series expansions are all based on the function $\exp(z)$. All three of these functions, by their very definitions, have meaning only for real values x. Nevertheless, the power series appearing in (1)–(3) converge absolutely for all real x, and so they converge absolutely if we replace the real variable x by the complex variable z. That is, we can define the following functions of the complex variable z:

$$e^z = 1 + \frac{z}{1!} + \frac{z^2}{2!} + \cdots + \frac{z^n}{n!} + \cdots, \tag{4}$$

$$\cos z = 1 - \frac{z^2}{2!} + \cdots + (-1)^n \frac{z^{2n}}{(2n)!} + \cdots, \tag{5}$$

$$\sin z = z - \frac{z^3}{3!} + \cdots + (-1)^n \frac{z^{2n+1}}{(2n+1)!} + \cdots. \tag{6}$$

In this way we extend functions defined only for real values to complex values. We do this of course by using power series expansions for the original functions of a real variable. One may ask if the extensions (4)–(6) are in some way accidental. Could other power series extensions exist? In §20, we will show that any power series extension of functions defined for real values to complex values is unique.

From the definition (1) in §18 and from (4), we see that

$$e^z = \exp(z). \tag{7}$$

Thus the function e^z has the following properties:

$$e^0 = 1, \tag{8}$$

$$e^{z+w} = e^z e^w, \tag{9}$$

$$e^{-z} = \frac{1}{e^z}, \tag{10}$$

$$e^{w-z} = \frac{e^w}{e^z} \tag{11}$$

(see §18, (3), (4), (8), and (9)). Furthermore, the function e^z is continuous, since $\exp(z)$ is continuous. The functions $\cos z$ and $\sin z$ are new functions of a complex variable, but they are closely connected with the function $e^z = \exp(z)$. Indeed, in the series (4), replace z by iz and compare the result with the series (5) and (6). We arrive at the identity

$$e^{iz} = \cos z + i \sin z. \tag{12}$$

This formula does not follow from the corresponding formula (20) of §18, which we proved for real values of the argument. Here of course we have proved it for all complex values. From (5) and (6) we see that

$$\cos(-z) = \cos z \quad \text{and} \quad \sin(-z) = -\sin z.$$

That is, $\cos z$ is an even function and $\sin z$ is an odd function. From these identities and (12), we find that

$$e^{-iz} = \cos z - i \sin z. \tag{13}$$

Solving (12) and (13) for $\sin z$ and $\cos z$, we find that

$$\cos z = \frac{e^{iz} + e^{-iz}}{2} \tag{14}$$

and

$$\sin z = \frac{e^{iz} - e^{-iz}}{2}. \tag{15}$$

Thus our new functions $\cos z$ and $\sin z$ can easily be expressed in terms of the function e^z.

The formulas for the sine of the sum of two numbers and for the cosine of the sum of two numbers must be proved anew for the functions $\cos z$ and $\sin z$. This can be done easily using (14), (15), and (9). I leave this to the reader. These formulas show the likenesses of the functions $\cos z$ and $\sin z$ and

the corresponding functions of a real variable. There is an important difference: for complex values, these functions are unbounded.

With formula (9), we can make a more detailed study of the function e^z
of the complex variable $z = x + i y$. Indeed, we have (see §18, (20))

$$e^{x+iy} = e^x e^{iy} = e^x(\cos y + i\sin y). \tag{16}$$

The first factor $\rho = e^x$ is the exponential function. It is fairly easy to
understand. It is a monotone increasing function which goes to infinity with
great rapidity as $x \to \infty$ and goes to zero, also with great rapidity, as $x \to -\infty$. It is positive for all values of x. We sketch its graph in Fig. 62. It intersects the axis of ordinates at the point $\rho = 1$.

The inverse function for the function e^x is the solution of the equation

$$\rho = e^x$$

with respect to x. It is called *the logarithm of ρ to the base e* or *the natural
logarithm of ρ*. We write it as $\ln \rho$, that is, $x = \ln \rho$. The real-valued function
$\ln \rho$ is defined only for positive values of ρ. It is positive for $\rho > 1$ and is
negative for $0 < \rho < 1$. We obtain the graph of the natural logarithm by reflecting the graph of the function $\rho = e^x$ about the line that bisects the first
and third quadrants of the coordinate plane. We sketch the graph in Fig. 63.

The functions $\cos y$ and $\sin y$ have period 2π. Therefore the function e^z
has the imaginary period $2\pi i$, as the identity (16) shows:

$$e^{z+2\pi i} = e^z$$

for all complex numbers z. Formula (16) also shows that the function e^z
vanishes for no value of z:

$$e^z \neq 0.$$

(This also follows at once from the identity (10).)

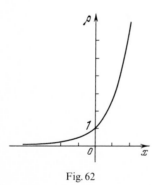

Fig. 62

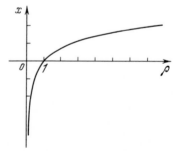

Fig. 63

The Function ln w of the Complex Variable w. To study the function $\ln w$ of the complex variable w, we must solve the equation

$$e^z = w = \rho(\cos \varphi + i \sin \varphi)$$

with respect to z. We rewrite our equation in the form

$$e^{x+iy} = e^x(\cos y + i \sin y) = \rho(\cos \varphi + i \sin \varphi).$$

We find that

$$\left.\begin{aligned} x &= \ln \rho, \\ y &= \varphi + 2k\pi. \end{aligned}\right\} \tag{17}$$

In formula (17), we refer to the real logarithm of the positive real number ρ, and k can be an arbitrary integer. That is, the complex function $\ln w$ of the complex variable w is a many-valued function. It is defined only up to the additive constant $2k\pi i$. This is *a priori* obvious, since the function e^z has the imaginary period $2\pi i$. The natural logarithm of a positive number is also defined only up to a constant $2k\pi i$, but as a rule we consider only its real part, which is defined uniquely. We are compelled here to extend the concept of a function: a function may be many-valued. Even in elementary algebra this difficulty is encountered when we extract roots, and in elementary trigonometry when we consider the inverse trigonometric functions.

To get an idea of how the many-valuedness of $\ln w$ manifests itself, let us suppose that the point w moves in the plane of the complex variable w, tracing out some curve that does not pass through the origin. This curve is described parametrically by an equation

$$w = w(t),$$

where the real parameter t may be thought of as time, increasing from 0. If

$$w = \rho(\cos \varphi + i \sin \varphi),$$

then $w(t) = [\rho(t), \varphi(t)]$, where $\rho(t)$ and $\varphi(t)$ are the polar coordinates of the point $w(t)$. Since the point $w(t)$ never goes through the origin in the course of its wanderings, the function $\rho(t)$ is positive and depends continuously on t. The function $\varphi(t)$, on the other hand, is not uniquely defined. To define it uniquely, we have to give it some determined value at the initial moment $t = 0$. That is, we make a choice of the value $\varphi(0)$, and from then on we suppose that $\varphi(t)$ varies continuously with t. The point w may return at some positive time t_1 to its starting point. That is, we may have

$$\rho(t_1) = \rho(0) \quad \text{and} \quad \varphi(t_1) = \varphi(0) + 2k\pi.$$

Here k may be different from 0.

Thus the value of the function $\ln w$ will vary from its initial value after $w(t)$ has described a closed curve not going through the origin, by an amount $2k\pi i$. It is easy to see that the integer k is the number of times that the point $w(t)$ winds around the origin in the course of its meanderings. This is made perfectly clear for the motion described by

$$w(t) = \cos t + i \sin t,$$

where t_1 has the value $2k\pi$. In this case, the point $w(t)$ circles the origin k times along the circle with radius 1 and center at the origin, and the logarithm changes from its original value by adding $2k\pi i$. This behavior of the logarithm. To find the function $\arccos w$, we must solve the equation $w = \cos z$ with respect to z. To do this, we use the formula (14) for the cosine:

The Functions $\arccos w$ and $\arcsin w$ of the Complex Variable w. The functions $\arccos w$ and $\arcsin w$ are easily expressed in terms of the complex logarithm. To find the function $\arccos w$, we must solve the equation $w = \cos z$ with respect to z. To do this, we use the formula (14) for the cosine:

$$w = \frac{e^{iz} + e^{-iz}}{2}.$$

We write

$$e^{iz} = \xi, \tag{18}$$

and obtain

$$\xi^2 - 2w\xi + 1 = 0,$$

so that $\xi = w + \sqrt{w^2 - 1}$. Here we interpret $\sqrt{w^2 - 1}$ as a *two-valued* function. and so we do not need to write both signs before the square root symbol. From (18) and the definition of the complex logarithm, we thus have

$$z = -i\ln(w + \sqrt{w^2 - 1}),$$

which is to say that

$$\arccos w = -i\ln(w + \sqrt{w^2 - 1}).$$

In much the same way, we find that

$$\arcsin w = -i\ln(iw + \sqrt{1 - w^2}).$$

Thus we can express the functions $\arccos w$ and $\arcsin w$ in terms of logarithms. The two-valued function $\sqrt{}$ and $\ln$ may assume all possible values.

Example. **The History of Logarithms.** The concept of a logarithm is well known in elementary mathematics. It originally was conceived of by comparing properties of geometric and arithmetic progressions:

$$\ldots, q^{-n}, \ldots, q^{-1}, 1, q, \ldots, q^{n}, \ldots, \tag{19}$$

and

$$\ldots, -n\alpha, \ldots, -\alpha, 0, \alpha, \ldots, n\alpha, \ldots, \tag{20}$$

where $q > 1$ and $\alpha > 0$. It is clear that multiplying numbers in the first row corresponds to adding numbers in the second row. In modern language, we say that the numbers of the second row are the logarithms of the numbers of the first row to a certain base. This connections between the rows (19) and (20) was partly known to Archimedes and was very familiar to mathematicians of the 15th and 16th centuries. However, this connection was not exploited for practical purposes. The "tables" (19) and (20) make it possible to replace multiplication by the simpler operation of addition. Later in the 16th century, a variety of computational methods, in particular the use of tables of logarithms, were developed to meet the needs of navigation. This was connected with the art of determining a ship's position by the stars. The first tables of logarithms were constructed by the Scottish mathematician John Napier (1550–1617) and later by the Swiss mathematician Louis Bourguet (1678–1742). Bourguet's fundamental idea was the following.

In constructing a table of logarithms, it is important that the numbers in the sequence (19) be sufficiently dense, which is to say that the number q should be close to 1. One chooses $q = 1 + \frac{1}{k}$, where k is a sufficiently large positive integer. In the sequence (20), one then chooses $\alpha = \frac{1}{k}$. With these choices, the sequences (19) and (20) are

$$\ldots, \left(1+\frac{1}{k}\right)^{-n}, \ldots, \left(1+\frac{1}{k}\right)^{-1}, 1, \left(1+\frac{1}{k}\right), \ldots, \left(1+\frac{1}{k}\right)^{n}, \ldots \tag{21}$$

and

$$\ldots, -\frac{n}{k}, \ldots, -\frac{1}{k}, 0, \frac{1}{k}, \ldots, \frac{n}{k}, \ldots \tag{22}$$

respectively. Then the numbers of the sequence (22) are the logarithms, term by term, of the numbers of the sequence (21), to the base $\left(1+\frac{1}{k}\right)^{k}$. The larger k is, the denser is the sequence (21) and the better is the table of logarithms constructed from the sequences (21) and (22). Bourguet took 10,000 for k and for the base of logarithms $\left(1+\frac{1}{k}\right)^{k}$ got the value. 2.71824, which agrees with the number e to four places of decimals. As we proved in Example 1 of §18, the number e is in fact the limiting value of $\left(1+\frac{1}{k}\right)^{k}$ as $k \to \infty$. When one goes to the limit in this fashion, the correspondence between the sequences (21) and (22) becomes the logarithmic relation $y = \ln x$.

§20. Power Series

In §18, we defined the function $\exp(z)$ of the complex variable z as the sum of a certain power series, $viz.$,

$$\exp(z) = \frac{1}{0!} + \frac{z}{1!} + \frac{z^2}{2!} + \cdots + \frac{z^n}{n!} + \cdots. \tag{1}$$

It is natural to generalize this construction by considering an arbitrary function $f(z)$ of the complex variable z that can be represented in the form of some power series

$$f(z) = a_0 + a_1 z + a_2 z^2 + \cdots + a_n z^n + \cdots, \tag{2}$$

where the numbers $a_0, a_1, a_2, \ldots, a_n, \ldots$ are complex constants that define the series in (1) and hence the function $f(z)$. The series (1) is plainly a special case of (2), with the choice

$$a_n = \frac{1}{n!}.$$

Plainly the function $f(z)$ is defined by (2) only for z's such that the series in (2) converges. As we showed in §18, the series in (1) converges for all complex numbers z. An arbitrary series (2) need not, of course, enjoy this property. The first problem we encounter, therefore, is the problem of determining the values of z for which a specific series as in (2) converges, or, which is even more important, converges absolutely.

Plainly the series in (2) converges for $z = 0$. There exist series that converge *only* for $z = 0$. We give such a series in Example 1 below. Such series, which represent no interesting function, will be designated as *trivial*.

Some Properties of Nontrivial Series. We start with the hypothesis that the series (2) converges for some value $z = z_0 \neq 0$. In view of (5) in §17, we have

$$\lim_{n \to \infty} a_n z_0^n = 0,$$

and so in particular all of the numbers

$$a_n z_0^n \quad \text{for } n = 0, 1, 2, \ldots$$

are bounded in absolute value by some constant a: $|a_n| \cdot |z_0|^n < a$.

Thus we have an estimate for the coefficients a_n:

$$|a_n| < \frac{a}{|z_0|^n} \quad \text{for} \ \ n = 0, 1, 2, \dots .$$

We now construct an auxiliary series with positive terms:

$$\hat{f}(z) = a + a\frac{|z|}{|z_0|} + a\frac{|z|^2}{|z_0|^2} + \cdots + a\frac{|z|^n}{|z_0|^n} + \cdots . \tag{3}$$

(The series (3) is not a power series.) The series (3) *majorizes* the series (2) in the sense of the comparison test stated in §17:

$$|a_n z^n| < a\frac{|z|^n}{|z_0|^n} \quad \text{for} \ \ n = 0, 1, 2, \dots . \tag{4}$$

For all values of z such that

$$|z| < |z_0|, \tag{5}$$

the series (3) is a convergent geometric progression. Therefore the series (2) is absolutely convergent for all z satisfying the inequality (5): this is simply the comparison test once again.

In the series (2), let us drop the constant term. We obtain the series

$$f_1(z) = a_1 z + a_2 z^2 + \cdots + a_n z^n + \cdots .$$

This series is majorized by the series

$$\hat{f}_1(z) = a\frac{|z|}{|z_0|} + a\frac{|z|^2}{|z_0|^2} + \cdots + a\frac{|z|^n}{|z_0|^n} + \cdots . \tag{6}$$

We therefore have the inequality

$$|f_1(z)| < \hat{f}_1(z).$$

For z satisfying (5), the series (6) is a geometric progression with sum

$$\hat{f}_1(z) = \frac{a\dfrac{|z|}{|z_0|}}{1 - \dfrac{|z|}{|z_0|}} = \frac{a|z|}{|z_0| - |z|}$$

(see formula (21) in §17). It follows that

$$|f_1(z)| < \frac{a|z|}{|z_0| - |z|} . \tag{7}$$

For our original function $f(z)$, we have

$$f(z)=a_0+f_1(z), \tag{8}$$

and $f_1(z)$ satisfies the inequality (7). We suppose that

$$a_0 \neq 0.$$

Let us now take $|z|$ so small that the modulus of the first summand in (8) is larger than the modulus of the second summand. Then the sum (8), which is to say the function $f(z)$, is different from 0. It suffices to take z such that

$$|z|<\frac{|a_0|}{2a}|z_0|=\rho. \tag{9}$$

Replace $|z|$ in (7) by its majorant (9). Doing this, we only increase the right side of (7), and so we find that

$$|f_1(z)|<\frac{a\dfrac{|a_0|}{2a}|z_0|}{|z_0|-\dfrac{|a_0|}{2a}|z_0|}=\frac{a\,|a_0|}{2a-|a_0|}<\frac{a\,|a_0|}{a}=|a_0|.$$

Thus, if $a_0 \neq 0$, there is a positive number ρ (see (9)) such that if

$$|z|<\rho,$$

the series (2) converges to a nonzero sum $f(z)$:

$$f(z)\neq 0 \quad \text{for } |z|<\rho. \tag{10}$$

The Radius of Convergence. Let (2) be nontrivial. There exists a positive number r (which may be equal to ∞) such that the series (2) converges absolutely for $|z|<r$ and diverges for $|z|>r$. The number r is called *the radius of convergence of the series* (2), and the open disk with center at the point 0 and radius r is called *the disk of convergence* of the series.

We will prove that the radius of convergence exists. Let M be the set of all real numbers of the form $|z_0|$, where z_0 is a complex number such that

$$\text{the series (2) converges for } z=z_0. \tag{11}$$

The condition (11) and the construction of the preceding paragraph show that if α is a positive number in M and $0<\beta<\alpha$, then β also belongs to M. Suppose first that M contains arbitrarily large numbers. Then M consists of all nonnegative real numbers, and in this case we define r as ∞. Suppose next that M does not contain arbitrarily large numbers, that is, M is bounded. In this case, we define r as the supremum of the set M. In both cases, M contains all numbers α such that $0 < \alpha < r$. We will prove that our number r is the radius of convergence of the series (2).

Suppose that $|z_1| < r$. There is then a number z_0 such that the series (2) converges for $z = z_0$ and

$$|z_1| < |z_0| < r.$$

Condition (5) shows that the series (2) converges absolutely for $z = z_1$. Therefore the series (2) converges absolutely for all points z such that $|z| < r$. On the other hand, suppose that $|z_1| > r$. Then by definition z_1 does not belong to M and so the series (2) diverges for $z = z_1$. This proves the existence of a positive radius of convergence for every nontrivial series (2). If the series (2) converges only for $z = 0$, we will say that it has radius of convergence zero, and from now on we will not use the term "nontrivial series".

The Behavior of a Power Series Near the Point $z = 0$. Let (2) be a power series with positive radius of convergence for which at least one of the coefficients a_n is different from 0. We will prove that there is a positive number ρ such that the equation

$$f(z) = 0 \tag{12}$$

has, in the disk $|z| < \rho$, either no solutions at all or the unique solution $z = 0$.

Let a_k be the first coefficient in (2) that is different from 0. We have already dealt with the case $k = 0$ (see (10)). In this case, the equation (12) has no solutions at all for sufficiently small positive ρ. Suppose now that $k > 0$. We write the function $f(z)$ in the form

$$f(z) = z^k f_k(z),$$

where

$$f_k(z) = a_k + a_{k+1} z^{k+1} + \cdots + a_{k+n} z^{k+n} + \cdots. \tag{13}$$

The constant term a_k in the series (13) is different from 0. By (10), there is a positive number ρ_k such that $f_k(z) \neq 0$ for $|z| < \rho_k$. It follows that $f(z)$ vanishes for $|z| < \rho_k$ only where $z^k = 0$, which is to say, only for $z = 0$.

The Uniqueness Theorem for Power Series. Let M be an infinite set of points in the plane of the complex variable z having 0 as a limit point (see §12, Example 3). We suppose that the function $f(z)$, defined by the power series

$$f(z) = a_0 + a_1 z + \cdots + a_n z^n + \cdots, \tag{14}$$

vanishes everywhere on M:

$$f(z) = 0 \quad \text{for all } z \in M. \tag{15}$$

Then all of the coefficients a_n of the series (14) are equal to 0.

Let us prove this fact. Assume that not all coefficients of (14) are equal to 0. Then, as proved above, there is a positive number ρ such that the equation $f(z)=0$ has in the disk $|z|<\rho$ not more than one solution, which is the point 0 if it exists at all. The inequality $|z|<\rho$ defines the neighborhood U_ρ of radius ρ of the point 0. Since the point $z=0$ is a limit point of the set M, by hypothesis the neighborhood U_ρ must contain points of M different from 0. The condition (15) states that $f(z)$ vanishes for nonzero points within U, and this is impossible. We must therefore infer that all of the coefficients of the series (14) are zero.

We draw a simple consequence of the result just proved. Suppose that we have two power series

$$\varphi(z)=\alpha_0+\alpha_1 z+\cdots+\alpha_n z^n+\cdots \tag{16}$$

and

$$\psi(z)=\beta_0+\beta_1 z+\cdots+\beta_n z^n+\cdots, \tag{17}$$

both with positive radius of convergence. Again let M be an infinite subset of the complex plane having 0 as a limit point. Suppose that

$$\varphi(z)=\psi(z) \tag{18}$$

for all points z in M. Then the coefficients of the series (16) and (17) are the same:

$$\alpha_n=\beta_n \quad \text{for } n=0,1,2,\dots. \tag{19}$$

To prove this, consider the difference of the functions $\varphi(z)$ and $\varphi(z)$:

$$f(z)=\varphi(z)-\psi(z)=(\alpha_0-\beta_0)+(\alpha_1-\beta_1)z+\cdots+(\alpha_n-\beta_n)z^n+\cdots. \tag{20}$$

The equalities (18) show that $f(z)=0$ for all $z\in M$. Therefore all of the coefficients of the series (20) vanish, which is to say that (19) holds.

Representing functions by means of power series is one of the most important techniques of higher mathematics. The use of power series presents us with a number of questions, among them the following. It is possible for a function $f(x)$ of the real variable x to be represented by two different power series,

$$\varphi(x)=\alpha_0+\alpha_1 x+\cdots+\alpha_n x^n+\cdots \tag{21}$$

and

$$\varphi(x)=\beta_0+\beta_1 x+\cdots+\beta_n x^n+\cdots, \tag{22}$$

on some tiny interval, say for $|x|<\rho$, where ρ is some positive number? The uniqueness theorem immediately gives us a negative answer to this question: the coefficients of the series (21) and (22) must be identical if they yield the same function for x in some interval centered at 0.

The argument to prove this is as follows. In (21) and (22), replace the real variable x by the complex variable z. We obtain complex functions $\varphi(z)$ and $\psi(z)$ which are equal for the points in the set M that consists of all real numbers x such that $|x| < \rho$. The point 0 is certainly a limit point of this set. The uniqueness theorem shows that the coefficients of the series (21) and (22) must be the same.

At the same time we answer a different question. Let $f(x)$ be a function of the real variable x that can be represented by a power series on some interval $|x| < \rho$. This function can be extended in only one way to a function of a complex variable that is represented by a complex power series. This is the case, for example, with the functions e^x, $\cos x$, and $\sin x$ defined for all real numbers x. In the preceding section, we extended these functions to all complex numbers z. It is now clear that such a power series extension is unique.

We now compute the radii of convergence of some well-known power series.

Example 1. We first give a power series with radius of convergence 0:

$$1 + 1! z + 2! z^2 + \cdots + n! z^n + \cdots . \tag{23}$$

To see that this series diverges for every $z \neq 0$, write $z = \dfrac{1}{\alpha}$. Thus we have

$$n! z^n = \frac{n!}{\alpha^n},$$

and, as proved in §13, (18), we have

$$\lim_{n \to \infty} \frac{n!}{\alpha^n} = \infty.$$

Thus the terms of the series (23) not only fail to go to zero, but are unbounded. Consequently the series (23) diverges for $z \neq 0$.

Example 2. Consider the series

$$f(z) = 1 + z + z^2 + \cdots + z^n + \cdots . \tag{24}$$

This is a geometric progression. For $|z| < 1$ it converges, and its sum is given by formula (21) of §17:

$$f(z) = \frac{1}{1-z}. \tag{25}$$

For $|z| \geq 1$, the series (24) diverges because the terms do not have limit 0. Thus the radius of convergence of (24) is 1. We encounter with this example a very important as well as interesting fact. The series (24) defines the function $f(z)$ only for $|z| < 1$. At the same time, the identity (25) defines the function $f(z)$ for all $z \neq 1$. This phenomenon is of great importance, and is typical in the theory of analytic functions. At the beginning one defines an analytic function by means of a power series, which of course specifies the function only within the disk of convergence. It may well happen that the given function is actually defined also for other values of the argument. In the case before us, the function is defined for all $z \neq 1$. Such extension of the domain of definition of a function, which is originally defined only within a disk of convergence, is called *analytic continuation* of the original function. Such continuation, it turns out, can be carried out in only one way. It is also typical of analytic functions that there is a singularity of the function $f(z)$ defined by (25) on the boundary of the disk of convergence of the series (24). In this case, the singularity occurs at the point $z = 1$, where the function defined by (25) becomes infinite or is simply undefined. Suppose that we begin with the function defined by (25) and seek to expand it in a power series. The disk of convergence of the power series can certainly not contain the point 1, since the function defined by (25) becomes infinite at 1. Therefore a power series expansion for the function (25) cannot have radius of convergence greater than 1. We have already seen that the power series expansion of the function (25) has radius of convergence equal to 1. Thus the existence of a singular point of the function (25) on the circle $|z| = 1$ is the obstacle to extending the disk of convergence.

Example 3. Let k be an integer. A natural generalization of the series (24) is the series

$$f_k(z) = 1 + 1^k z + 2^k z^2 + \cdots + n^k z^n + \cdots. \tag{26}$$

For $k = 0$, we have the series (24). Applying formulas (15) and (16) of §13, we see that the radius of convergence of the series (26) is equal to 1 for all values of k.

For $k = -1$, we get a series worthy of special attention. In §24, Example 3, we will show that its sum is $1 - \ln(1 - z)$. Thus we have

$$1 - \ln(1 - z) = 1 + \frac{z}{1} + \frac{z^2}{2} + \cdots + \frac{z^n}{n} + \cdots.$$

In this equality, we may replace z by $-z$ and obtain

$$\ln(1 + z) = \frac{z}{1} - \frac{z^2}{2} + \frac{z^3}{3} - \cdots + (-1)^{n-1} \frac{z^n}{n} + \cdots. \tag{27}$$

For $z = 1$, the series (27) converges, since it is a series with alternating signs and terms decreasing to 0 in absolute value (see §16). For $z = -1$, the series diverges (see §16, (14)). Thus the series (27) converges at some points of the boundary of the disk of convergence and diverges at others. For $k = -2$, we obtain the series

$$f_{-2}(z) = 1 + \frac{z}{1^2} + \frac{z^2}{2^2} + \cdots + \frac{z^n}{n^2} + \cdots.$$

This series converges absolutely at all points of the boundary of the disk of convergence, that is, for all z such that $|z| = 1$. (See Example 2 of §17.)

These examples show that the behavior of a power series on the boundary of the disk of convergence varies considerably. The power series $f_0(z)$ diverges at all of these points; the power series $f_1(z)$ converges at some and diverges at others; and the power series $f_2(z)$ converges at all of them.

Example 4. Let z_0 be any fixed complex number. We may define a function $f(z)$ by a power series

$$f(z) = a_0 + a_1(z - z_0) + \cdots + a_n(z - z_0)^n + \cdots. \tag{28}$$

Here of course the powers are the powers of $z - z_0$, not of z. We may repeat our previous arguments to show that there is a number $r \geq 0$ such that the series (28) converges absolutely for $|z - z_0| < r$ and diverges for $|z - z_0| > r$.

Chapter V
The Differential Calculus

In this chapter we give an extremely compact treatment of the first fundamental part of mathematical analysis. This is the concept of the derivative of a function, computing the derivatives of various functions, which is to say, differentiating them. To define the derivative, we must slightly alter the definition of limit as given in Chapter IV. We then give the definition of the derivative simultaneously for the real and complex cases. This occupies §21. In §22, we work out the derivatives of a number of functions. We first show how to differentiate a power series, and this gives us the means of differentiating the functions e^z, $\cos z$, and $\sin z$, power series expansions for which were obtained in Chapter IV We then give the rules for differentiating products, quotients, and composite functions. The rule for differentiating an inverse function is derived from this last. Thus in §22 we obtain the derivatives of all of the elementary transcendental functions. Although the present chapter bears the title "The differential calculus", we avail ourselves of the opportunity in it of defining integration as the inverse of differentiation. We define the primitive function, or indefinite integral, again for the real and complex cases simultaneously. We use Lagrange's formula to prove the uniqueness of the primitive function up to an additive constant. Hence we must first prove Rolle's theorem and Lagrange's formula for real-valued functions. All of this is dealt with in §23. In §24, we find the primitives of a number of specific functions and also give the rules for integration by parts and by change of variable. We give in §25 a somewhat artificial definition of the definite integral as the particular primitive function that vanishes at a specified point z_0. This definition yields a number of rules relating to definite integrals. In particular, we give an estimate of a definite integral of a real-valued function in terms of the modulus of this function. We again use Lagrange's formula in our proof of this estimate.

We have introduced the definite integral in a way that is not wholly natural. We did this in order to give in §26 (the last of this chapter) Taylor's formula with the remainder in integral form, and also to give estimates in various cases for the size of the remainder. The derivation of the integral form for the remainder in Taylor's formula is carried out by a very simple argument. I have not encountered elsewhere so simple a proof for this version of Taylor's formula.

§21. The Derivative

What is called the derivative of a function plays an enormous rôle in mathematical analysis. Suppose that we are studying a certain function $f(t)$ of a real variable t. Using certain operations which we will describe in detail further on, we construct a new function $f'(t)$ from the function $f(t)$. This new function is called *the derivative of* $f(t)$. Before giving a precise definition of how we obtain $f'(t)$ from $f(t)$, we will say a few words about the function $f'(t)$. Our verbal description of the derivative is of course rather cloudy, but still it will give us some idea of the significance of the derivative for applications of mathematics, in particular to physics and to other branches of the natural sciences. For definiteness let us suppose that t is time, and that $f(t)$ is some sort of physical quantity that varies with time. Such quantities that vary with time are encountered in all physical processes. For example, the abscissa x of a point that moves on the axis of abscissas depends upon time and so is a function of time. We can describe this motion by

$$x = f(t).$$

Such an equation tells us how the position of the point varies with the passage of time. Similarly, the temperature T of a body may vary with time, and the process of this variation may be written as

$$T = f(t),$$

where $f(t)$ is a certain specific function of the time t. Without at first blush trying to define exactly what we mean, we can picture to ourselves that the speed with which the quantity $f(t)$ changes may itself vary with time. Thus we perceive clearly that the speed of change of $f(t)$ with time is itself a function of the time t. This speed is what we denote by $f'(t)$. A purely verbal definition of the derivative $f'(t)$ is: $f'(t)$ is the speed with which the quantity $f(t)$ is changing at time t.

In studying a physical process, we customarily list a number of physical quantities that characterize the state of our physical object at a given moment of time. These quantities typically vary with time, and so describe our physical process. Suppose for example that we are studying the compression of an amount of gas in a cylinder. The physical quantities that characterize the state of the gas at a given moment are the following: the volume of the gas; its pressure; and its temperature. If we compress the gas, all three of these quantities change with time. Plainly the speed with which these quantities change is essential in describing the process of compression of the gas (see §24, Example 1). It is an observed fact that all, or almost all, laws of physics can be stated mathematically in the form of equations connecting physical quantities with the speeds with which they vary. That is, these equations connect functions and their derivatives. This circumstance is the reason for the paramount significance of the derivative in studying physical processes.

In order to give a complete mathematical description of the derivative, we must first modify the operation of passage to the limit, which we studied in §12.

A New Concept of Passage to the Limit. In defining limits in §12, we began with an certain sequence of numbers

$$s_1, s_2, \ldots, s_n, \ldots. \tag{1}$$

We then explained what is meant by the assertion that the sequence (1) converges to a number s. The sequence (1) converges to the number s if and only if the difference $s_n - s$ is arbitrarily small for all sufficiently large values of the index n. When this happens, we write

$$\lim_{n \to \infty} s_n = s. \tag{2}$$

Let us repeat here the more formal exact description of the limit process adopted in mathematics, which we presented in §12. The relation (2) holds if and only if for every positive number ε, we can choose a positive integer v so large that the inequality $|s_n - s| < \varepsilon$ holds for all indices $n > v$.

We can paraphrase this definition of passage to a limit by taking as our initial object not a sequence of numbers but a function $f(n)$ of a positive integral argument, which is defined by the identity

$$f(n) = s_n \quad \text{for } n = 1, 2, \ldots. \tag{3}$$

It is clear that if we know the sequence (1), then we know the function $f(n)$ and conversely. The relation between sequences and functions of a positive integral argument is stated in (3).

It is now reasonable to consider not functions $f(n)$ of a positive integral argument but functions $f(z)$ of a real or complex argument z. In studying functions $f(n)$ we supposed that n increases without bound, or as we say, goes to infinity. In considering a function $f(z)$ of a real or complex argument z, we suppose that z converges to some numerical value z_0. We then ask, does the value $f(z)$ converge to some number f_0 as z converges to z_0? If this happens, we write

$$\lim_{z \to z_0} f(z) = f_0. \tag{4}$$

We note that the function $f(z)$ need not be defined for $z = z_0$.

The statement (4) has the following evident interpretation. The closer the argument z gets to the value z_0, so the closer and closer does $f(z)$ approach to the value f_0. The exact formulation now adopted in mathematics is as follows. The relation (4) is said to hold if and only if, for every positive number ε, one can choose a sufficiently small positive number δ so that for all z's such that $|z-z_0|<\delta$, the inequality $|f(z)-f_0|<\delta$ holds.

This new definition of passage to a limit is not only analogous to the old definition for sequences, but is connected logically with it. However, we will not explore this logical connection.

Continuity of a Function. With our new definition of limit for functions, we can give a precise definition of continuity of a function. A function $f(z)$ is said to be *continuous at a point* z_0 where it is defined if the relation

$$\lim_{z \to z_0} f(z) = f(z_0) \tag{5}$$

holds. This new definition of continuity is equivalent to the old definition that we gave in Example 1, §12, p. 120. In the sequel we will use only the new definition. In Example 3, p. 192, we will prove that the two definitions are in fact equivalent.

The Derivative of a Function. In mathematics we define derivatives both for functions of a real argument and for functions of a complex argument. The definitions are the same for both cases. Hence we begin with a function $f(z)$, where z denotes either a real or a complex variable. We fix a value z of the argument of the function $f(z)$ and will define the number $f'(z)$, the derivative of $f(z)$, at this fixed point z. Along with the fixed value z, we consider an arbitrary value ζ of the argument of the function $f(z)$. We think of ζ as obtained from z by changing z. To emphasize this viewpoint, we define the quantity

$$\Delta z = \zeta - z$$

as the *increment of the quantity* z. Thus we have

$$\zeta = z + \Delta z.$$

Given the increment Δz of the independent variable z, we define the corresponding increment of the function $f(z)$:

$$\Delta f(z) = f(\zeta) - f(z) = f(z + \Delta z) - f(z).$$

To define the derivative $f'(z)$, we first compute the ratio of the increment of $f(z)$ to the increment of the argument:

$$\frac{\Delta f(z)}{\Delta z} = \frac{f(\zeta) - f(z)}{\zeta - z} = \frac{f(z + \Delta z) - f(z)}{\Delta z}. \tag{6}$$

This difference quotient is defined for all values of ζ different from z, which is to say for all $\Delta z \neq 0$. For $\zeta = z$, both numerator and denominator of the quotient (6) vanish and so the quotient is not defined. Suppose now that the difference quotient (6) converges to some limit as $\zeta \to z$ (equivalently, as $\Delta z \to 0$). That is, suppose that

$$\lim_{\zeta \to z} \frac{f(\zeta) - f(z)}{\zeta - z} \tag{7}$$

exists. We then say that the function $f(z)$ *has a derivative $f'(z)$ at the point z.* We define the derivative $f'(z)$ as the limit in (7):

$$f'(z) = \lim_{\zeta \to z} \frac{f(\zeta) - f(z)}{\zeta - z}. \tag{8}$$

In defining the derivative of $f(z)$ at the point z, we will always suppose that all values of ζ that are sufficiently close to z are points where the function is defined. We will also adopt the convention that whenever we write a derivative, it exists.

Sometimes a function $f(z)$ is presented to us as some complicated expression. It may then be convenient to write the symbol "'" (read "prime") that indicates the derivative outside of parentheses:

$$f'(z) = (f(z))'. \tag{9}$$

We make a final comment about the limit in (7) and (8). As already remarked, we suppose that the function f is defined in some neighborhood U_δ of the point z (see §12, (12) and (16)). Therefore in computing the limit (7), we may choose an arbitrary number ζ (real or complex, depending upon the case we are studying) satisfying the inequalities $0 < |\zeta - z| < \delta$.

Infinitely Small Quantities. Along with the new definition of limit that we have presented in this section, we present a new notion of infinitely small quantities. Suppose that we have some variable quantity τ that is of interest to us mainly as it goes to zero. We will then call it *an infinitely small quantity.* Thus the increment $\Delta z = \zeta - z$ of the independent variable is infinitely small in the computation of a derivative. If τ is an infinitely small quantity and $\varphi(\tau)$ is a function of τ such that

$$\lim_{\tau \to 0} \varphi(\tau) = 0,$$

then we will also say that the quantity $\varphi(\tau)$ is infinitely small. More precisely, we say that the quantity $\varphi(\tau)$ is infinitely small relative to τ. Very often we need to compare the rapidity of decrease of the quantity $\varphi(\tau)$ with the rapidity of decrease of τ itself. We do this with the use of the symbols O and o (see §13). Thus, if the quantity

$$\frac{\varphi(\tau)}{\tau} \tag{10}$$

remains bounded as $\tau \to 0$, we write

$$\varphi(\tau) = O(\tau).$$

If the quotient (10) goes to 0 as $\tau \to 0$, we write

$$\varphi(\tau) = o(\tau).$$

Frequently we need to compare the rapidity of decrease of a quantity $\varphi(\tau)$ with the rapidity of decrease of the quantity τ^n, where n is some positive integer. If we have

$$\varphi(\tau) = O(\tau^n),$$

we say that $\varphi(\tau)$ is *infinitely small of order n with respect to* τ.

The Differential. Let us analyze the definition of the derivative in a bit more detail. As in (8), we have

$$f'(z) = \lim_{\zeta \to z} \frac{f(\zeta) - f(z)}{\zeta - z}.$$

In other words, the quantity

$$\alpha(\zeta, z) = \frac{f(\zeta) - f(z)}{\zeta - z} - f'(z)$$

is infinitely small with respect to $\zeta - z$, and the last equality can be rewritten as

$$f(\zeta) - f(z) = (f'(z) + \alpha)(\zeta - z), \tag{11}$$

where

$$\lim_{\zeta \to z} \alpha = 0.$$

Let us write

$$\alpha \cdot (\zeta - z) = \beta.$$

Then we rewrite (11) in the form

$$f(\zeta) - f(z) = f'(z)(\zeta - z) + \beta, \tag{12}$$

where evidently

$$\beta = o(\zeta - z). \tag{13}$$

The left side of the equality (12) is the increment of the function, while the right side consists of two terms, the second of which satisfies (13). Therefore if $f'(z) \neq 0$, the second term of the right side of (12) goes to 0 faster than the first term. We thus consider that the first term of the right side of (12) is the principal term. Thus we may say that the principal part of the increment $\Delta f(z)$ of the function $f(z)$ is equal to $f'(z)\Delta z$. We now introduce some new terminology. We will call the increment Δz of the independent variable *the differential of z* and we will write

$$\Delta z = dz.$$

We will call the principal part of the increment $\Delta f(z)$ *the differential of $f(z)$* and will denote it by $df(z)$. Thus we have the formal identity

$$df(z) = f'(z)\,dz. \tag{14}$$

At first glance, the concept of differential is almost vacuous. It would appear to give us only a new notation for the derivative, as (14) can be rewritten in the form

$$f'(z) = \frac{df(z)}{dz}. \tag{15}$$

However, as time goes on we will convince ourselves that the employment of differentials is useful. We sometimes write the derivative $f'(z)$ in the form

$$f'(z) = \frac{d}{dz} f(z). \tag{16}$$

The notations (15) and (16) for the derivative offer some advantages over the notation $f'(z)$. For one thing, these notations make clear what is the variable with respect to which the derivative is being taken. In some cases this can be of considerable importance. The operation of forming the derivative is often called *the operation of differentiation:* this is suggested by the notation (16). A function that has a derivative at a point z is said to be *differentiable at the point z.* A function that has a derivative for all values of its argument is called *a differentiable function.*

Continuity of a Differentiable Function. If a function $f(z)$ has a derivative $f'(z)$ at a point z, then it is continuous at z. Going to the limit in the equality (12) as $\zeta \to z$, we find that

$$\lim_{\zeta \to z}(f(\zeta) - f(z)) = \lim_{\zeta \to z}(f'(z)(\zeta - z) + o(\zeta - z)) = 0.$$

By (5), therefore, the function $f(z)$ is continuous at the point z. Also, if a function has a derivative at every point where it is defined, it is continuous.

Partial Derivatives. In our discussion of differentiation so far, we have confined our attention to functions of a single variable. In reality one is mostly concerned with functions of several variables. For example, a polynomial in a variable z is also a function of all of its coefficients. In differen-

tiating a polynomial we are of course well aware that we should differentiate with respect to the variable z. Nevertheless, it frequently occurs that we
have to do with a function of (say) two variables, $f(z, w)$, and it may be essential to show with respect to which of the two variables we are to differentiate. Such a derivative is called *a partial derivative* and is denoted by the
symbol

$$\frac{\partial}{\partial z} f(z, w). \tag{17}$$

In computing the partial derivative (17) we give an increment only to z and
hold w constant.

Example 1. We now give a geometric interpretation of the derivative of a
real-valued function $f(x)$ of the real variable x. We sketch the graph K of
the function $f(x)$ in a plane P, in which we have chosen a rectangular Cartesian coordinate system. See Fig. 64. Thus K is the curve defined by the
equation

$$y = f(x). \tag{18}$$

We fix a certain value x of the argument of our function, and denote by y
the corresponding value of the function, defined as in (18). Thus $a = (x, y)$ is
a point of the plane P that lies on the curve K. We think of this point as
fixed. Along with the value x, we consider any other value $\xi \neq x$, and write η
for the corresponding value of the function, that is, $\eta = f(\xi)$. The point
$\alpha = (\xi, \eta)$ lies on the curve K. This point we will think of as movable. Now
suppose that ξ approaches x while decreasing. Then the point α moves
along the curve K and approaches the point a from the right. Suppose on
the other hand that ξ approaches x while increasing. Then the point α approaches the point a along the curve K from the left. We draw the line L
that passes through the points a and α. We say that L is a *secant of the
curve K*. We compute the slope of the line L, that is, the tangent of the
angle between L and the positive axis of abscissas. To compute this slope,
we draw a horizontal line through the point a, that is, a line parallel to the

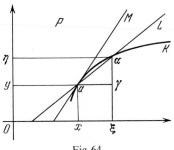

Fig. 64

axis of abscissas. Through the point α we draw a vertical line. Let γ be the point of intersection of these two lines. We must compute the tangent of the angle at the vertex a in the triangle whose vertices are $a\gamma\alpha$, taking into account the sign of the angle. We consider the specific case $\xi > x$. Then the positive number $\xi - x$ is equal to the length of the side $a\gamma$ of our triangle, and the number $\eta - y$ is the length of the side $\gamma\alpha$ if $\eta - y$ is positive, and the number $-(\eta - y)$ is the length of the side if $\eta - y$ is negative. In both cases the slope of the line L is the quotient $\dfrac{\eta - y}{\xi - x}$. If $\xi < x$, the slope of the line L is equal to the same quotient. If we take note of the signs of both quantities $\eta - y$ and $\xi - x$, this is easy to see. Now suppose that $f(x)$ has a derivative at the point x. This means that the quotient $\dfrac{\eta - y}{\xi - x}$ converges to the limit $f'(x)$ as $\xi \to x$. That is, if the derivative $f'(x)$ exists, the slope of the secant L converges to a definite limit as the point ξ converges to the point a. In this case, the line L also approaches a limiting position M. The line M passes through the point a. We define M as *the tangent line to the curve K at the point a*. Thus the derivative $f'(x)$ is the slope of the tangent line M to the curve K drawn at the point $a = (x, y)$, which of course lies on K. This is the geometric meaning of the derivative.

It can happen that the quotient $\dfrac{\eta - y}{\xi - x}$ converges to a limit, say k_1, as $\xi \to x$ through values greater than x, and that this quotient converges to a different limit, say k_2, as $\xi \to x$ through values less than x. Our definition of derivative shows that $f(x)$ then does not admit a derivative at x. The numbers k_1 and k_2 are called the *right* and *left derivatives*, respectively, of $f(x)$ at the point x. Let M_1 and M_2 be the limiting positions of the secants to the curve K as α approaches a from the right and left sides, respectively. Since $k_1 \neq k_2$, the lines M_1 and M_2 are distinct, and the curve K admits no tangent line at the point a. In this case, we call the point a *an angular point of the curve K*. The simplest example of an angular point is provided by the function

$$f(x) = |x|.$$

Its graph, described by the equation $y = |x|$, consists of the bisectrix of the first quadrant of the coordinate system together with the bisectrix of the second quadrant. For $x = 0$, we have $k_1 = +1$ and $k_2 = -1$, and the term "angular point" in this example acquires a clear geometric meaning.

Example 2. We give an important application of the notion of the derivative. Suppose that $f(x)$ is a real-valued function of a real variable x and that for some fixed value x, the derivative $f'(x)$ exists and is positive. We will show that in some neighborhood of x, the function is strictly increasing. More specifically, we will show that there is a positive number δ (possibly small) such that for $|\xi - x| < \delta$, the following conditions hold:

$$\text{if } \xi < x, \quad \text{then } f(\xi) < f(x);$$
$$\text{if } \xi > x, \quad \text{then } f(\xi) > f(x).$$

These assertions follow at once from formula (12). In just the same way, we prove that if $f'(x) < 0$, then $f(x)$ is strictly decreasing in the vicinity of x. That is, there is a positive number δ (possibly small) such that for $|\xi - x| < \delta$, the following conditions hold:

$$\text{if } \xi < x, \quad \text{then } f(\xi) > f(x);$$
$$\text{if } \xi > x, \quad \text{then } f(\xi) < f(x).$$

The last two results have a highly important consequence. Suppose that $f(x)$ attains its maximum value at some point x. That is, the inequality

$$f(x) \geq f(\xi)$$

holds for all ξ for which $f(\xi)$ is defined. If $f'(x)$ exists, then the value $f'(x)$ must be zero.

One can prove a stronger result. Suppose that the function $f(x)$ has a *local maximum* at a point x. That is, there is a positive number δ such that the inequality

$$f(x) \geq f(\xi) \tag{19}$$

holds for all ξ such that $|\xi - x| < \delta$. If $f'(x)$ exists, then we must have

$$f'(x) = 0.$$

We prove this by contradiction. Assume that $f'(x) > 0$. Then $f(x)$ is strictly increasing at x, so there is a value $\xi > x$ of the argument of $f(x)$ such that $\xi - x < \delta$ and such that

$$f(\xi) > f(x).$$

This contradicts our hypothesis (19). We see in the same way that $f'(x)$ cannot be negative. For, if it were, there would be a value $\xi < x$ satisfying the conditions $|\xi - x| < \delta$ and also

$$f(\xi) > f(x).$$

This too contradicts (19). Therefore the derivative of a function at a local maximum must be zero if it exists at all.

Similar statements hold for the point where a function assumes its minimum value and for points where a function assumes a local minimum value. That is, if we have $f(x) \leq f(\xi)$ for all ξ such that $|\xi - x| < \delta$, then the derivative of $f(x)$ at x must be zero, if it exists at all.

Thus to find local maxima and local minima of a differentiable function, we must solve the equation

$$f'(x)=0$$

and then examine in more detail the behavior of $f(x)$ near the points where $f'(x)=0$. In ordinary parlance, local maxima and local minima are simply called maxima and minima, respectively.

Example 3. We now prove the equivalence of our two definitions of continuity. Suppose that $f(z)$ is continuous at a point z_0 in the sense of the present section. That is, suppose that for every positive number ε, there exists a positive number δ such that

$$\text{for } |z-z_0|<\delta, \quad \text{we have } |f(z)-f(z_0)|<\varepsilon. \tag{20}$$

We wish to prove that $f(z)$ is continuous at the point z_0 also in the sense of Example 1 of §12. That is, for every sequence

$$z_1, z_2, \ldots, z_n, \ldots$$

of points where $f(z)$ is defined, the condition

$$\lim_{n\to\infty} z_n = z_0 \tag{21}$$

implies that

$$\lim_{n\to\infty} f(z_n) = f(z_0). \tag{22}$$

We choose an arbitrary positive number ε, and then define δ so that (20) holds. The condition (21) implies that there exists a positive integer v such that the inequality

$$|z_n - z_0| < \delta$$

holds for all $n > v$. Then condition (20) shows that

$$|f(z_n) - f(z_0)| < \varepsilon. \tag{23}$$

Thus, for every positive number ε, there exists a positive integer v such that (23) holds for all $n > v$. This means that condition (22) holds.

In the sequel, we shall take continuity in the sense of the present section.[1]

---

[1] *Translator's note.* The converse of Example 3 is also true. If a function is sequentially continuous at a point z_0 then it is continuous at z_0 in the sense of the present section. This is most simply proved by contradiction. That is, suppose that $f(z)$ is not continous at z_0 in the sense of the present section. Then there is a positive number α such that the inequality

$$|f(z)-f(z_0)|\geq\alpha \tag{*}$$

holds for points z that are arbitrarily close to (but of course distinct from) z_0. Choose z_1 to be any point for which (*) holds and $|z_1-z_0|<1$. When $z_1, z_2, \ldots, z_n$ have been chosen, let z_{n+1} be a point for which (*) holds and the number $|z_{n+1}-z_0|$ is less than $\dfrac{1}{n+1}$ and also less than all of the numbers $|z_1-z_0|, \ldots, |z_n-z_0|$. Then $z_1, z_2, \ldots, z_n \ldots$ is a sequence of distinct points with limit z_0 for which (22) fails. That is, if $f(z)$ is sequentially continuous at z_0, it must also be continuous at z_0 in the sense of the present section.

Example 4. At the end of §12, we set down five basic rules for limits of sequences. Exactly the same rules hold for limits in the sense of the present section. We will set them down here.

Let $f(z)$ and $g(z)$ be two functions of a real or complex variable for which we have

$$\lim_{z \to z_0} f(z) = f_0 \quad \text{and} \quad \lim_{z \to z_0} g(z) = g_0.$$

Then the following hold:

Rule 1. $\lim\limits_{z \to z_0} (f(z) + g(z)) = f_0 + g_0;$

Rule 2. $\lim\limits_{z \to z_0} (f(z) - g(z)) = f_0 - g_0;$

Rule 3. $\lim\limits_{z \to z_0} (f(z) \cdot g(z)) = f_0 g_0;$

Rule 4. if $g_0 \neq 0$, then

$$\lim_{z \to z_0} \frac{f(z)}{g(z)} = \frac{f_0}{g_0};$$

Rule 5. if $f(z)$ and $g(z)$ are real-valued functions and $f(z) \leq g(z)$ for all z, then $f_0 \leq g_0$.

To establish these rules, we proceed as in §12, using infinitely small quantities. Thus we set

$$f(z) = f_0 + \hat{f}(z) \quad \text{and} \quad g(z) = g_0 + \hat{g}(z).$$

It is clear that $\hat{f}(z)$ and $\hat{g}(z)$ are infinitely small quantities with respect to the quantity $z - z_0$. Now translate the obvious and simple properties of infinitely small quantities with respect to sequences that we stated at the end of §12 on page 118 into analogous properties for infinitely small quantities in the present context. Proofs of the limit rules 1 through 5 are easily obtained in this way.

§22. Computing Derivatives

We will now compute the derivatives of a number of functions, beginning with polynomials and convergent power series. The latter will give us at once the derivatives of the functions e^z, $\cos z$, and $\sin z$. We then give the rules for differentiating products and quotients of two functions. Next we show how to differentiate a composite function, that is, a function of the form $f(z) = \psi(\varphi(z))$, and how to differentiate an inverse function. Using the last rule, we will differentiate the inverse trigonometric functions and the logarithm.

We begin by setting down some of the simplest rules for differentiating.

We note first that the derivative of a constant function is zero. For, if $f(z) = c$, then we have $f(\zeta) = c$, so that $f(\zeta) - f(z) = 0$, which is to say that

$$c' = 0. \tag{1}$$

Now let $f_1(z), f_2(z), \ldots, f_n(z)$ be functions and let $\alpha_1, \alpha_2, \ldots, \alpha_n$ be constants. We find that

$$(\alpha_1 f_1(z) + \alpha_2 f_2(z) + \cdots + \alpha_n f_n(z))'$$
$$= \alpha_1 f_1'(z) + \alpha_2 f_2'(z) + \cdots + \alpha_n f_n'(z). \tag{2}$$

To prove (2) it suffices to prove two simpler identities:

$$(f(z) + g(z))' = f'(z) + g'(z) \tag{3}$$

and

$$(\alpha f(z))' = \alpha f'(z) \tag{4}$$

We use the rules for limits given in Example 4 of §21. Let us prove (3). We have

$$(\alpha f(z) + g(z))' = \lim_{\zeta \to z} \frac{f(\zeta) + g(\zeta) - (f(z) + g(z))}{\zeta - z}$$
$$= \lim_{\zeta \to z} \left(\frac{f(\zeta) - f(z)}{\zeta - z} + \frac{g(\zeta) - g(z)}{\zeta - z} \right)$$
$$= \lim_{\zeta \to z} \frac{f(\zeta) - f(z)}{\zeta - z} + \lim_{\zeta \to z} \frac{g(\zeta) - g(z)}{\zeta - z}$$
$$= f'(z) + g'(z).$$

The proof of (4) is just as easy. We have

$$(\alpha f(z))' = \lim_{\zeta \to z} \frac{\alpha f(\zeta) - \alpha f(z)}{\zeta - z} = \alpha \lim_{\zeta \to z} \frac{f(\zeta) - f(z)}{\zeta - z} = f'(z).$$

Let us now use (3) and (4) to prove (2). We use induction. For $n = 1$, (2) is (4). Suppose that (2) holds when the number of summands is $n - 1$. Write the left side of (2) as the sum of two summands:

$$((\alpha_1 f_1(z) + \alpha_2 f_2(z) + \cdots + \alpha_{n-1} f_{n-1}(z)) + \alpha_n f_n(z))'.$$

In view of (3) and (4), this derivative is equal to

$$(\alpha_1 f_1(z) + \alpha_2 f(z) + \cdots + \alpha_{n-1} f_{n-1}(z))' + \alpha_n f_n'(z). \tag{5}$$

Our inductive hypothesis now shows that the expression (5) is equal to the right side of the identity (2). This proves (2).

Let us apply (2) to a polynomial

$$a_0 + a_1 z + \cdots + a_n z^n. \tag{6}$$

We find that

$$(a_0 + a_1 z + a_2 z^2 + \cdots + a_n z^n)' = a_0' + a_1(z)' + a_2(z^2)' + \cdots + a_n(z^n)'.$$

Taking account of (1), we see that in order to differentiate the polynomial (6), we need only compute the derivative of the monomial functions z^k for positive integers k. We will prove that

$$(z^k)' = k z^{k-1}. \tag{7}$$

We use formula (11) of §17 to write

$$(z^k)' = \lim_{\zeta \to z} \frac{\zeta^k - z^k}{\zeta - z} = \lim_{\zeta \to z} (\zeta^{k-1} + \zeta^{k-2} z + \cdots + z^{k-1}) = k z^{k-1}.$$

(Take note again of the limit rules given on p. 193.) Thus we find that the derivative of the polynomial (6) is

$$(a_0 + a_1 z + a_2 z^2 + \cdots + a_n z^n)' = a_1 + 2a_2 z + 3a_3 z^2 + \cdots + na_n z^{n-1}. \tag{8}$$

Note that the identity (8) is valid for polynomials in both real and complex arguments.

The Derivative of a Power Series. Let the function $f(z)$ be defined by the power series

$$f(z) = a_0 + a_1 z + a_2 z^2 + \cdots + a_n z^n + \cdots \tag{9}$$

with positive radius of convergence r. We will prove that $f'(z)$ exists and find what it is. Indeed, we define

$$f_1(z) = a_1 + 2a_2 z + 3a_3 z^2 + \cdots + na_n z^{n-1} + \cdots. \tag{10}$$

It follows from formula (17) of §13 and formula (4) of §20 that the power series $f_1(z)$ has radius of convergence equal to r. We will prove that

$$f'(z) = f_1(z) \tag{11}$$

for all z such that $|z| < r$. We do this by a direct computation and estimation of the difference quotient

$$\frac{f(\zeta) - f(z)}{\zeta - z}$$

for each fixed z such that $|z| < r$. Write s for the number $\frac{1}{2}(r + |z|)$, noting that $|z| < s < r$. Since we are concerned with the limit of the difference quotient as $\zeta \to z$, we may suppose that $|\zeta| < s$ for all ζ that we deal with.

We set down three ancillary series of positive terms:

$$\hat{f}_0 = |a_0| + |a_1|s + |a_2|s^2 + \cdots + |a_n|s^n + \cdots;$$
$$\hat{f}_1 = |a_1| + 2|a_2|s + 3|a_3|s^3 + \cdots + n|a_n|s^{n-1} + \cdots;$$
$$\hat{f}_2 = 2^2|a_2| + 3^2|a_3|s + 4^2|a_3|s^2 + \cdots + n^2|a_n|s^{n-2} + \cdots.$$

To see that these series converge, apply the two formulas cited in the preceding paragraph.

Making use of Example 3 in §16, we write the difference quotient

$$\frac{f(\zeta) - f(z)}{\zeta - z}$$

in the form of a series:

$$\frac{f(\zeta) - f(z)}{\zeta - z} = \frac{a_0 - a_0}{\zeta - z} + a_1 \frac{\zeta - z}{\zeta - z} + \cdots + a_n \frac{(\zeta^n - z^n)}{\zeta - z} + \cdots. \tag{12}$$

Apply formula (11) of §17 to find that

$$\left| a_n \frac{(\zeta^n - z^n)}{\zeta - z} \right| = |a_n| \cdot |\zeta^{n-1} + \zeta^{n-2}z + \cdots + z^{n-1}| \leq |a_n| n s^{n-1}.$$

Hence the series $\hat{f}_1$ majorizes the series (12) term by term, and so the comparison test shows that the series (12) converges. Now subtract the series (10) from the series (12). We find

$$\frac{f(\zeta) - f(z)}{\zeta - z} - f_1(z) = a_2 \left(\frac{\zeta^2 - z^2}{\zeta - z} - 2z \right) + \cdots + a_n \left(\frac{\zeta^n - z^n}{\zeta - z} - n z^{n-1} \right) + \cdots. \tag{13}$$

We make an estimate of the quantity $\dfrac{\zeta^n - z^n}{\zeta - z} - n z^{n-1}$:

$$\left| \frac{\zeta^n - z^n}{\zeta - z} - n z^{n-1} \right| = |(\zeta^{n-1} - z^{n-1}) + z(\zeta^{n-2} - z^{n-2}) + \cdots$$
$$+ z^{n-2}(\zeta - z) + (z^{n-1} - z^{n-1})|$$
$$= |\zeta - z| |(\zeta^{n-2} + \zeta^{n-3}z + \cdots + z^{n-2})$$
$$+ z(\zeta^{n-3} + \zeta^{n-4}z + \cdots + z^{n-3}) + \cdots + z^{n-2}|$$
$$\leq |\zeta - z| n^2 s^{n-2}.$$

Hence the series $\hat{f}_2$ majorizes the series (13) term by term, when the terms of $\hat{f}_2$ are multiplied by $|\zeta - z|$. We see finally that

$$\left| \frac{f(\zeta) - f(z)}{\zeta - z} - f_1(z) \right| < |\zeta - z| c,$$

where c is a positive constant. It follows that

$$\lim_{\zeta \to z} \frac{f(\zeta) - f(z)}{\zeta - z}$$

exists and is equal to $f_1(z)$. This proves (11). In words: power series may be differentiated term by term within their open disks of convergence.

The identity (11) and the series (10) applied to some familiar power series give us at once the derivatives

$$(e^z)' = e^z \tag{14}$$

and

$$\sin' z = \cos z, \qquad \cos' z = -\sin z. \tag{15}$$

Here of course we use the power series expansions given in (1)–(3) of §19. These series converge for all complex numbers z. Therefore the identities (14) and (15) hold for all complex and all real values of z.

We turn now to some fundamental formulas of differentiation. We will compute the derivatives of products, quotients, and composite functions.

The Derivative of a Product. Let $f(z)$ and $g(z)$ be two functions. We have

$$(f(z) \cdot g(z))' = f'(z) \cdot g(z) + f(z) \cdot g'(z). \tag{16}$$

To prove this, we write the difference $f(\zeta) \cdot g(\zeta) - f(z) \cdot g(z)$ in a useful form:

$$f(\zeta) \cdot g(\zeta) - f(z) \cdot g(z) = f(\zeta) \cdot g(\zeta) - f(z) \cdot g(\zeta) + f(z) \cdot g(\zeta) - f(z) g(z)$$
$$= (f(\zeta) - f(z)) g(\zeta) + f(z)(g(\zeta) - g(z)).$$

Dividing both sides of this identity by $\zeta - z$, we find

$$\frac{f(\zeta) \cdot g(\zeta) - f(z) \cdot g(z)}{\zeta - z} = \frac{f(\zeta) - f(z)}{\zeta - z} g(\zeta) + f(z) \frac{g(\zeta) - g(z)}{\zeta - z}.$$

Take the limit as $\zeta \to z$ in this identity and obtain (16).

The Derivative of a Quotient. Let $f(z)$ and $g(z)$ be two functions and suppose that $g(z) \neq 0$. We have

$$\left[\left(\frac{f(z)}{g(z)} \right)' \right] = \frac{f'(z) g(z) - g'(z) f(z)}{(g(z))^2}. \tag{17}$$

To prove this, we first write the increment of $\dfrac{f(z)}{g(z)}$:

$$\frac{f(\zeta)}{g(\zeta)} - \frac{f(z)}{g(z)} = \frac{f(\zeta)\,g(z) - f(z)\,g(\zeta)}{g(\zeta)\,g(z)}. \tag{18}$$

We rewrite the numerator of the right side of (18):

$$f(\zeta)\,g(z) - f(z)\,g(\zeta) = f(\zeta)\,g(z) - f(z)\,g(z) + f(z)\,g(z) - g(\zeta)\,f(z)$$
$$= (f(\zeta) - f(z))\,g(z) - f(z)(g(\zeta) - g(z)). \tag{19}$$

Divide both sides of (18) by $\zeta - z$ and use the identity (19). Taking the limit of the result as $\zeta \to z$, we obtain (17).

As a result of (17), (15), and the identity

$$\sin^2 z + \cos^2 z = 1$$

(which can be proved by multiplying power series and adding), we find the derivatives of

$$\tan z = \frac{\sin z}{\cos z} \quad \text{and} \quad \operatorname{ctn} z = \frac{1}{\tan z}.$$

They are

$$\tan' z = \frac{1}{\cos^2 z} \quad \text{and} \quad \operatorname{ctn}' z = -\frac{1}{\sin^2 z}. \tag{20}$$

The Derivative of a Composite Function $f(z) = \psi(\varphi(z))$. Let $\varphi(z)$ and $\psi(w)$ be two functions. In the second function, let us substitute for w the expression $w = \varphi(z)$. We obtain in this way what is called *the composite function* $\psi(\varphi(z))$, or as one also says, *a function of a function*. We will prove that

$$f'(z) = \psi'(w)\,\varphi'(z) = \psi'(\varphi(z))\,\varphi'(z). \tag{21}$$

Let us write the derivative as the quotient of differentials, as in §21, (15). Then (21) assumes the following easily remembered form:

$$\frac{d\psi(w)}{dz} = \frac{d\psi(w)}{dw}\frac{dw}{dz}, \tag{22}$$

where $w = \varphi(z)$.

To prove (21), we begin by writing $\varphi(\zeta) = \omega$. We then have

$$\psi(\varphi(\zeta)) - \psi(\varphi(z)) = (\psi'(w) + \alpha)(\omega - w),$$

arcsin z, arccos z, arctan z, and arcctn z. We find the following results:

$$\ln' z = \frac{1}{z};\tag{27}$$

$$\arcsin' z = \frac{1}{\sqrt{1-z^2}} \quad \text{and} \quad \arccos' z = \frac{-1}{\sqrt{1-z^2}};\tag{28}$$

$$\arctan' z = \frac{1}{1+z^2} \quad \text{and} \quad \text{arcctn}' z = \frac{-1}{1+z^2}.\tag{29}$$

Recall that the function $\ln z$ is defined as a solution of the equation $e^w = z$ with respect to w. Formula (25) shows that

$$\ln' z = \frac{1}{(e^w)'} = \frac{1}{e^w} = \frac{1}{z}.$$

We prove only the first of the two identities (28). The function arcsin z is defined as a solution of the equation $\sin w = z$ with respect to w. The formula (25) in this case yields

$$\arcsin' z = \frac{1}{\sin' w} = \frac{1}{\cos w} = \frac{1}{\sqrt{1-\sin^2 w}} = \frac{1}{\sqrt{1-z^2}}.$$

We also prove only the first of the identities (29). We have

$$\arctan' z = \frac{1}{\tan' w} = \cos^2 w = \frac{1}{1+\tan^2 w} = \frac{1}{1+z^2}.$$

Finally let us compute the derivative of the function z^r, where r is an arbitrary real number. For a rational number r and positive real z, this number is defined in elementary algebra. For arbitrary real r and complex z, we define z^r by

$$z^r = e^{r \ln z}.\tag{30}$$

We differentiate (30) by using the rule for differentiating a composite function and also using (27). We thus obtain

$$(z^r)' = \frac{de^w}{dw} \cdot \frac{dw}{dz} = e^w \cdot \frac{dw}{dz} = e^w \cdot r \cdot \frac{1}{z} = e^{r \ln z} \cdot r \cdot \frac{1}{z} = r z^r \cdot \frac{1}{z} = r z^{r-1}.$$

That is, we have

$$(z^r)' = r z^{r-1}.\tag{31}$$

As we stated in §19, the elementary transcendental functions are the functions e^z, $\sin z$, $\cos z$, $\tan z$, $\text{ctn} z$, and their inverse functions. The class of where $\lim\limits_{\omega \to w} \alpha = 0$ (see (11) in §21). Dividing this identity by $\zeta - z$, we obtain

$$\frac{f(\zeta) - f(z)}{\zeta - z} = (\psi'(w) + \alpha)\left(\frac{\omega - w}{\zeta - z}\right). \tag{23}$$

Take the limit as $\zeta \to z$ in (23) and obtain (21), inasmuch as $\lim\limits_{\zeta \to z}(\omega - w) = 0$.

The Derivative of an Inverse Function. Suppose that we can solve the equation $\psi(w) = z$ for w as a function of z. That is, suppose that we can find a continuous function $w = \varphi(z)$ for which the identity

$$\psi(\varphi(z)) = z \tag{24}$$

obtains. Then the function $\varphi(z)$ is called *the inverse function to* $w = \psi(z)$. (We have already encountered some special cases in §19.) Suppose furthermore that at a fixed point w_0 the nonequality $\psi'(w_0) \neq 0$ holds. Then the function $\varphi(z)$ is differentiable at the point $z_0 = \psi(w_0)$ and its derivative is given by the formula

$$\varphi'(z) = \frac{1}{\psi'(w)} = \frac{1}{\psi'(\varphi(z))} \tag{25}$$

in a neighborhood of the point z_0.

To prove this, we use (24) to write

$$(\zeta - z) = \psi(\varphi(\zeta)) - \psi(\varphi(z)) = (\psi'(\varphi(z)) + \alpha)(\varphi(\zeta) - \varphi(z)): \tag{26}$$

again see formula (11) of §21. Note that $\lim \alpha = 0$ as $\varphi(\zeta) \to \varphi(z)$, which is to say, as $\zeta \to z$. Now suppose that $\psi'(\varphi(z_0)) \neq 0$. Then we have $\psi'(\varphi(z_0)) + \alpha \neq 0$ for ζ sufficiently close to z_0. We infer from (26) that

$$\frac{\varphi(\zeta) - \varphi(z_0)}{\zeta - z_0} = \frac{1}{\psi'(\varphi(z_0)) + \alpha}.$$

In this identity, take the limit as $\zeta \to z_0$. This gives the identity (25).

The identity (25) holds for both real and complex variables z.

Suppose that the function $z = \psi(w)$ that we considered in the preceding discussion is expansible in a power series in the variable $w - w_0$, having positive radius of convergence. Suppose also that $\psi'(w_0) \neq 0$. We will prove in §35 that the inverse function $\varphi(z)$ exists in some neighborhood of the point $z_0 = \psi(w_0)$ and is expansible in a power series with positive radius of convergence in the variable $z - z_0$. Naturally we have $w_0 = \varphi(z_0)$.

Let us now compute the derivatives of the functions inverse to e^w, $\sin w$, $\cos w$, $\tan w$, and $\text{ctn } w$. These are the functions that we have written as $\ln z$, *algebraic functions* consists of all polynomials, rational functions (that is, quotients of polynomials), and functions obtainable from rational functions by the extraction of roots. We now extend the definition of elementary transcendental function to include all functions that can be obtained from the ten functions just listed and from algebraic functions by composing these functions with each other an arbitrary finite number of times and in an arbitrary order. With the rules given in this section, we can differentiate all elementary transcendental functions.

Example 1. We now give an interpretation of the derivative in terms of elementary mechanics. Suppose that we have a point moving on the axis of abscissas, so that its abscissa is a specific function of time:

$$x = f(t). \tag{32}$$

If the point moves with constant velocity, the equality (32) is

$$x = x_0 + vt,$$

where x_0 is the initial position of the point at $t = 0$, and v is its velocity. If the point moves but not with constant velocity, one must explain what is meant by its velocity at a given moment. We do this in mechanics as follows. Along with the moment t, we consider another moment $\tau > t$. During the time from t to τ, the point moves from position $f(t)$ to position $f(\tau)$. It is reasonable to define the average velocity of the point during the time from t to τ as the quotient $\dfrac{f(\tau) - f(t)}{\tau - t}$. We then define the velocity of the point at the moment t as the limit of this quotient, which is to say the derivative of the function $f(t)$. That is, we define

$$v(t) = f'(t).$$

Suppose now that the velocity $v(t)$ is not constant. We can then consider acceleration of the point. The average acceleration during the time from t to τ is defined as the quotient $\dfrac{v(\tau) - v(t)}{\tau - t}$. The acceleration $u(t)$ at the moment t is defined as the limit of this quotient as $\tau \to t$. That is, we define

$$u(t) = v'(t).$$

Thus the acceleration is the derivative of the derivative $f'(t)$. It is called *the second derivative of the function* $f(t)$ and is denoted by the symbol $f''(t)$. Thus the acceleration of the point is defined by

$$u(t) = f''(t).$$

§23. The Indefinite Integral

In §§21 and 22, we have defined and discussed derivatives of functions $h(z)$ (it is convenient now to write a generic function as $h(z)$) of both real and complex variables. That is, for such a function $h(z)$, we formed a second function $h'(z)$, the derivative of $h(z)$. We recall from §21, (15), a second notation for the derivative:

$$h'(z) = \frac{dh(z)}{dz}.$$

Recall as well that the operation of going from the function $h(z)$ to the function

$$f(z) = h'(z) \tag{1}$$

is called *the operation of taking the derivative of the function $h(z)$* or *the operation of differentiating the function $h(z)$*.

Whenever mathematicians study an operation of any sort, the problem always arises of finding the inverse operation. In algebra, for example, when one considers multiplication, one also considers the inverse operation of division; with raising numbers to powers one considers extraction of roots; and so on. In speaking of the inverse to a given operation, we must always consider two extremely important questions. First, does an inverse operation exist? Second, is the inverse operation uniquely defined? For example, division is not always possible, since division by zero is impossible. Extraction of roots is not always uniquely defined. For example, every positive number admits two distinct square roots.

The operation inverse to the operation of differentiation is the operation of finding a function $h(z)$ when all we know is its derivative $f(z)$: $h'(z) = f(z)$, $f(z)$ being a known function. This operation is called *the operation of integration*. We use a special symbol for it:

$$h(z) = \int f(d)\,dz. \tag{2}$$

The symbol $\int$ is called *an integral sign*. Formula (2) is read as follows:

$h(z)$ is equal to the integral of $f(z)\,dz$.

The assertion that integration is the operation inverse to differentiation simply says that formulas (1) and (2) mean the same thing.

A function $h(z)$ that satisfies (1) is frequently called *a primitive of the function $f(z)$*. Thus integration is the process of finding primitives.

In the present section, we will give no general answer to the question of whether or not integration is always possible. We will show that a great many functions admit integrals that are easily computed. For example, it is clear that

$$\int \cos z \, dz = \sin z, \qquad \int \sin z \, dz = -\cos z. \tag{3}$$

These identities follow immediately from the formulas (15) of §22. In just the same way, it is easy to verify that

$$\int (\alpha_0 + \alpha_1 z + \cdots + \alpha_n z^n) \, dz = \alpha_0 z + \frac{\alpha_1}{2} z^2 + \frac{\alpha_2}{3} z^3 + \cdots + \frac{\alpha_n}{n+1} z^{n+1}. \tag{4}$$

This formula follows from formula (8) of §22. Thus, in some cases at least, the operation of integration is possible and is a perfectly reasonable topic for study.

We will give a complete answer in this section to the second question, namely, whether or not integrals are unique. It is first of all clear that integrals are not unique. For, suppose that (2) holds. Then we also have

$$\int f(z) \, dz = h(z) + c, \tag{5}$$

where c is an arbitrary constant. This follows from the fact that

$$(h(z) + c)' = h'(z) + c' = h'(z)$$

(see §22, identities (1) and (3)).

Therefore the integral is defined only up to an additive constant. It is for exactly this reason that the symbol

$$\int f(z) \, dz$$

is called *an indefinite integral*. We will prove in the present section that the indefinitess of the integral is limited to this additive constant. We will show in fact that if

$$\int f(z) \, dz = h_1(z) \quad \text{and} \quad \int f(z) \, dz = h_2(z), \tag{6}$$

then we have

$$h_2(z) = h_1(z) + c \tag{7}$$

where c is a constant. We remark that this assertion holds only under the restriction that the function $f(z)$ be defined on a connected set Ω, that is, the set of numbers z where $f(z)$ is defined forms a connected set Ω.

In the case of a function defined on a subset of the real numbers, connectedness of the set Ω means that if z_0 and z^* are two points where the function $f(z)$ is defined, say $z_0 < z^*$, then $f(z)$ is also defined for all z such that $z_0 < z < z^*$. That is, all such points z belong to the set Ω. In the case of a function defined for complex values, the set Ω is to be a connected subset of the complex plane, and in this situation, connectedness means something different. Suppose that z_0 and z^* are two points of the plane that belong to the set Ω where $f(z)$ is defined. Suppose that there is a broken line with vertices at

$$z_0, z_1, \ldots, z_{n-1}, z_n = z^* \tag{8}$$

that lies entirely within the set Ω. That is, we write $[z_j z_{j+1}]$ for the line segment in the plane with vertices at z_j and z_{j+1}. If all of the line segments $[z_j z_{j+1}]$ $(j=0,1,\ldots,n-1)$ are contained in Ω, we say that the broken line defined by (8) lies entirely within Ω. If such a broken line exists for every pair of points z_0 and z^* in Ω, we say that Ω is *connected*.

In order to prove that (6) implies (7), it suffices to show that the identity

$$h'(z) = 0 \tag{9}$$

implies that

$$h(z) = c, \tag{10}$$

where c is a constant. We suppose that the set Ω where $h(z)$ is defined is connected. Setting $h(z) = h_1(z) - h_2(z)$, we find that (10) implies (7).

We will give the proof that (9) implies (10) in a number of steps. The proof is by no means simple, and first we will prove some facts that are of importance in their own right.

The Maximum and Minimum of a Continuous Function. Let $h(x)$ be a real-valued continuous function defined on a closed interval

$$a \leqq x \leqq b \tag{11}$$

of the real number system. That is, all numbers x satisfying (11) may serve as arguments of the function $h(x)$. We consider no other numbers where $h(x)$ might be defined. The function $h(x)$ has a smallest value and a largest value as x runs through all the points (11). That is, there are values α and β of x as in (11) such that the inequalities

$$h(\alpha) \leqq h(x) \leqq h(\beta) \tag{12}$$

hold for all x such that $a \leqq x \leqq b$. In Example 2 below we will show that this property may fail for discontinuous functions. Thus the hypothesis that $h(x)$ be continuous is essential.

We will first prove that $h(x)$ is bounded. We use the method of contradiction. Assume that $h(x)$ is unbounded. Then for every natural number n, there is a value $x = \gamma_n$ of the argument for which the inequality

$$|h(\gamma_n)| > n \tag{13}$$

holds. The numbers γ_n all lie in the interval $a \leq x \leq b$. Therefore the sequence

$$\gamma_1, \gamma_2, \ldots, \gamma_n, \ldots \tag{14}$$

admits a subsequence

$$\delta_1 = \gamma_{n_1}, \; \delta_2 = \gamma_{n_2}, \ldots, \delta_k = \delta_{n_k}, \ldots \tag{15}$$

such that

$$\lim_{k \to \infty} \delta_k = \delta,$$

where δ is some number belonging to the closed interval $[ab]$ (see §15, Bounded Sequences, pp. 138–140, as well as §12, (18), p. 117). Since the function $h(x)$ is continuous, we have

$$\lim_{k \to \infty} h(\delta_k) = h(\delta):$$

but this is impossible. The sequence (15) is a subsequence of (14) and so, by (13), the numbers $|h(\delta_k)|$ get arbitrarily large as k increases. Hence they cannot approach the finite limit $h(\delta)$, and we have proved that $h(x)$ is bounded both above and below. Thus the set of all numbers $h(x)$ admits an infimum, which we will denote by p, and a supremum, which we will denote by q. As the supremum of the set of all numbers of the form $h(x)$, q is the limit of a sequence of numbers of the form $h(x)$. That is, there exists a sequence

$$\mu_1, \mu_2, \ldots, \mu_n, \ldots \tag{16}$$

of values of the argument x for which we have

$$\lim_{n \to \infty} h(\mu_n) = q. \tag{17}$$

All of the numbers in the sequence (16) belong to the interval $[ab]$, and so the sequence (16) admits a subsequence

$$v_1 = \mu_{n_1}, v_2 = \mu_{n_2}, \ldots, v_k = \mu_{n_k}, \ldots \tag{18}$$

that converges to a limit β:

$$\lim_{k \to \infty} v_k = \beta.$$

Since the sequence (18) is a subsequence of (16), we infer from (17) that

$$\lim_{k \to \infty} h(v_k) = h(\beta) = q.$$

Therefore the supremum of the set of all values of the function $h(x)$ is the value of the function at the point β. One proves in precisely the same way that the function $h(x)$ assumes its minimum value at a point α.

Corollary (Rolle's Theorem). Suppose now that the function $h(x)$ which we have been discussing has equal values at the endpoints a and b of the interval $[ab]$, $h(a)=h(b)$, and that $h(x)$ admits a derivative at every point x such that $a<x<b$. Then there is a point ξ in the interior of the interval $[ab]$ (that is, $a<\xi<b$) such that

$$h'(\xi)=0. \tag{19}$$

To prove this, suppose first that $h(x)$ is a constant. Then its derivative is equal to zero for all x such that $a<x<b$, and we can take any such point for the point ξ. Suppose now that the function $h(x)$ is nonconstant. There are two possibilities. First, there may be a point x such that $h(x)>h(a)$ $=h(b)$. If the first possibility fails, then there must be a point x such that $h(x)<h(a)=h(b)$. In the first case, the function $h(x)$ assumes its maximum value at some point ξ lying in the interior of $[ab]$. In the second case, it assumes its minimum value at some point ξ in the interior of $[ab]$. In both cases, the value of $h'(\xi)$ is zero (see Example 2 in §21). That is, (19) holds.

Finite Increments (Lagrange's Formula). Consider a real-valued continuous function $h(x)$ defined on a closed interval $[x_0\,x_1]$ of the real variable x. Suppose that $h(x)$ admits a derivative at every interior point of the interval $[x_0\,x_1]$. Then there is a point ξ in the interior of the interval $[x_0\,x_1]$ for which the equality

$$h(x_1)-h(x_0)=h'(\xi)(x_1-x_0) \tag{20}$$

holds.

To prove this, consider the function

$$g(x)=h(x)-\frac{h(x_1)-h(x_0)}{x_1-x_0}x. \tag{21}$$

The function $g(x)$ has equal values at x_0 and x_1:

$$g(x_1)-g(x_0)=h(x_1)-h(x_0)-\frac{h(x_1)-h(x_0)}{x_1-x_0}(x_1-x_0)=0.$$

Therefore Rolle's theorem is applicable to the function $g(x)$: there is a point ξ in the interior of $[x_0\,x_1]$ such that

$$g'(\xi)=0$$

(see (19)). In this equality we put the value for $g'(x)$ obtained from (21). This gives

$$h'(\xi)-\frac{h(x_1)-h(x_0)}{x_1-x_0}=0,$$

which gives us (20).

Lagrang's formula (20) provides an instant proof of our assertion for real-valued functions of a real variable, namely, the identity

$$h'(x) = 0 \tag{22}$$

(see (9)) implies that

$$h(x) = C$$

(see (10)). From (22) we see that the value $h'(\xi)$ in (20) is zero, and so $h(x_1) - h(x_0) = 0$ for any two points where $h(x)$ is defined. (The set where $h(x)$ is defined being connected, $h(x)$ is defined for all x such that $x_0 < x < x_1$.) Thus $h(x)$ is a constant.

We will now prove our fundamental fact for complex-valued functions $h(x)$ of a real variable. We write $h(x) = \varphi(x) + i\,\varphi(x)$, where $\varphi(x)$ and $\psi(x)$ are real-valued functions of the real variable x. By §22, (2), we see that $h'(x) = \varphi'(x) + i\psi'(x)$. Therefore, if $h'(x) = 0$, it follows that $\varphi'(x) = 0$ and $\psi'(x) = 0$. From what we have just proved, the functions $\varphi(x)$ and $\psi(x)$ must be constants, and so the function $h(x)$ is also a constant.

Proof of (10) for Functions of a Complex Variable. Consider first the case in which the broken line (8) consists of a single line segment $[z_0 z_1]$ in the complex plane. We introduce a real parameter t on this interval, writing

$$z = z_0(1 - t) + z_1 t. \tag{23}$$

as the real variable t increases from 0 to 1, the point z defined by (23) runs through the closed interval $[z_0 z_1]$. We may regard the function $h(z)$ on the closed interval $[z_0 z_1]$ as a function of the real variable t, by writing

$$\hat{h}(t) = h(z_0(1 - t) + z_1 t).$$

It follows from §22, formula (21) that

$$\hat{h}'(t) = h'(z_0(1 - t) + z_1 t) \cdot (z_1 - z_0).$$

Since the function $h'(z)$ is zero for all z in the closed interval $[z_0 z_1]$, the function $\hat{h}'(t)$ is equal to zero for all t in the interval $0 < t < 1$. Therefore we have $\hat{h}(0) = \hat{h}(1)$. Since $\hat{h}(0) = h(z_0)$ and $\hat{h}(1) = h(z_1)$, we see that $h(z_0) = h(z_1)$. Finally suppose that the broken line (8) consists of an arbitrary number of segments. By what we have just proved, we have the equalities

$$h(z_0) = h(z_1),$$
$$h(z_1) = h(z_2),$$

$$\cdots\cdots\cdots$$

$$h(z_{n-1}) = h(z_n) = h(z^*),$$

so that $h(z^*) = h(z_0)$. That is, $h(z)$ is a constant.

Functions of Several Variables. Suppose that the function under the integral sign (often called *the integrand*) depends not on a single variable, but on several. Thus it might depend upon two independent variables and so have the form $f(z, w)$. We must indicate clearly which of the variables is the variable with respect to which we are integrating. This is in fact made clear by the expression (2), where the symbol dz means that we are to integrate with respect to the variable z. Thus the integral

$$\int f(z, w)\, dz$$

itself is a function of the two variables z and w, and the relation (1) assumes the form

$$f(z, w) = \frac{\partial}{\partial z} h(z, w)$$

(compare with §21, (17)). In this case, the identity

$$\frac{\partial}{\partial z} h(z, w) = 0$$

implies that $h(z, w) = c(w)$, where $c(w)$ is an arbitrary function of the variable w. Instead of the relationship (7), we find that

$$h_2(z, w) = h_1(z, w) + c(w).$$

Example 1. Suppose that the set of all points z for which $h(z)$ is defined is disconnected. Then the derivative $h'(z)$ of the function $h(z)$ may be zero without the concomitant condition that $h(z)$ be constant. As an example, consider the function $\operatorname{sign} x$, defined for all nonzero real numbers x, by the conditions that $\operatorname{sign} x = 1$ for positive x and $\operatorname{sign} x = -1$ for negative x. The set where $\operatorname{sign} x$ is defined is disconnected, its derivative is zero everywhere, and it is nonconstant.

Example 2. We define a function $h(x)$ on the closed interval $0 \leq x \leq 1$ by the rules

$$h(0) = 0,$$

$$h(x) = 1 - x \qquad \text{for } x > 0.$$

Plainly this function is discontinuous at the point 0. Also it does not assume the least upper bound 1 of all of its values.

Example 3. Suppose that we have a point moving on the axis of abscissas with constant acceleration a (compare with Example 1 of §22). Let us find the equation of motion of this point. Let $x = f(t)$ be the coordinate of the point at time t. Thus we have the equation $f''(t) = a$. From this and formulas (4) and (7) we obtain

$$f'(t)=v_0+at,$$ (24)

where v_0 is the velocity of the point at the moment $t=0$. Apply formulas (4) and (7) once more to find that

$$f(t)=x_0+v_0 t+\tfrac{1}{2}at^2,$$

where x_0 is the position of the point at the moment $t=0$. This is the well-known formula for motion of a point with constant acceleration.

§24. Computation of Some Indefinite Integrals

We will give some rules for finding indefinite integrals. These formulas follow immediately from the formulas for differentiation that we gave in §22. Note that the right side of every formula contains an arbitrary additive constant c.

The Integral of a Power Series. The formula (10) of §22 for differentiating a power series gives us at once the formula for integrating a power series:

$$\int(\alpha_0+\alpha_1 z+\alpha_2 z^2+\cdots+\alpha_n z^n+\cdots)\,dz$$

$$=c+\alpha_0 z+\alpha_1\frac{z^2}{2}+\alpha_2\frac{z^3}{3}+\cdots+\alpha_n\frac{z^{n+1}}{n+1}+\cdots.$$ (1)

One should note that the radii of convergence of the two power series appearing in (1) are equal. Formula (1) is called *term by term integration for power series*.

The Integrals of Some Specific Functions. We list (with a little repetition) a few specific integrals. These integration formulas follow at once from formulas (14), (15), (20), (27), (28), and (29), respectively, in §22:

$$\int e^z\,dz=c+e^z;$$ (2)

$$\int\sin z\,dz=c-\cos z\quad\text{and}\quad\int\cos z\,dz=c+\sin z;$$ (3)

$$\int\frac{dz}{\cos^2 z}=c+\tan z\quad\text{and}\quad\int\frac{dz}{\sin^2 z}=c-\operatorname{ctn} z;$$ (4)

$$\int\frac{dz}{z}=c+\ln z;$$ (5)

$$\int\frac{dz}{\sqrt{1-z^2}}=c+\arcsin z;$$ (6)

$$\int\frac{dz}{1+z^2}=c+\arctan z.$$ (7)

Up to this point, we are able to write specific formulas for indefinite integrals only when the integrand is recognizable as the derivative of some function that we have already studied. Integration is thus a wholly different matter from differentiation. The rules set forth in §22 enable us to differentiate all elementary transcendental functions (including algebraic functions) and again obtain a function of the same class. The integral of an elementary transcendental function, however, need not be a function of this class. Integration, therefore, calls forth a whole new class of functions. Furthermore, even when the integral of some function is an elementary transcendental function, the task of actually finding the integral in closed form may be formidable. Despite this, there are some rules that enable us in many cases to transform an integrand into a function that is recognizable as the derivative of a known function. We now discuss two of these rules: 1) change of variable in the integrand; 2) integration by parts

Change of Variable in the Integrand. Suppose we know that a certain function $f(z)$ is the derivative of another function (or possibly the same function) $h(z)$:

$$f(z) = h'(z).$$

This of course means that

$$\int f(z)\,dz = h(z).$$

That is, suppose that we can integrate the function $f(z)$. Then we can make the change of variable $z = \psi(w)$ to generate a new function $g(w)$ that we can also integrate. That is, we write the composite function $h(\psi(w))$ and compute its derivative:

$$(h(\psi(w)))' = h'(\psi(w))\,\psi'(w) = f(\psi(w))\,\psi'(w).$$

Now define

$$g(w) = f(\psi(w)) \cdot \psi'(w). \tag{8}$$

We find at once that

$$\int g(w)\,dw = h(\psi(w)).$$

That is to say, if we can integrate the function $f(z)$, we can also integrate all functions of the form (8). Plainly we can obtain in return the function $f(z)$ by the substitution $w = \varphi(z)$, where $\varphi(z)$ is the inverse function of the function $\psi(w)$. The connection between the functions $f(z)$ and $g(w)$ is simple enough, but it becomes perfectly automatic if we make the substitution $z = \psi(w)$ both in the integrand $f(z)$ and in the differential dz. That is, we write

$$f(z)\,dz = f(\psi(w))\,d\psi(w) = f(\psi(w))\,\psi'(w)\,dw.$$

Compare this with formula (14) in §21.

The formal transition from one integrand with the variable z to a different integrand with variable w by the substitution $z = \psi(w)$ justifies our practice of writing the differential symbol dz next to the integrand. We illustrate this technique of integration with a simple example. Let $\varphi(z)$ be some function. We will find the integral

$$\int \frac{\varphi'(z)\,dz}{\varphi(z)}. \tag{9}$$

Make the substitution $\varphi(z) = w$. Then by our rules we can write the integral (9) as

$$\int \frac{\varphi'(z)\,dz}{\varphi(z)} = \int \frac{dw}{w} = c + \ln w = c + \ln \varphi(z). \tag{10}$$

Integration by Parts. The rule for differentiating a product is

$$(u(z)\,v(z))' = u'(z)\,v(z) + u(z)\,v'(z),$$

which is formula (16), §22. We interpret this in integral form as

$$\int u(z)\,v'(z)\,dz + \int v(z)\,u'(z)\,dz = c + u(z)\,v(z).$$

Writing

$$u'(z)\,dz = du(z) \quad \text{and} \quad v'(z)\,dz = dv(z),$$

we obtain the identities

$$\int u(z)\,dv(z) + \int v(z)\,du(z) = c + u(z)\,v(z)$$

and

$$\int u(z)\,dv(z) = c + u(z)\,v(z) - \int v(z)\,du(z). \tag{11}$$

Formula (11) is called *the formula for integration by parts*. We use it in cases where the integral on the right side can be computed, while the integral on the left is not immediately computable.

Example 1. We present an application of differentiation and integration to a problem in physics. We will study the behavior of an ideal gas confined in a container of variable volume, for example a cylinder with a movable piston. We suppose that the gas admits no exchange of heat with the surrounding medium, that the piston moves with no friction, and that the mass of the gas is one gram molecular weight. Let P be the pressure of the contained gas, V its volume, and T its temperature (absolute). The ideal gas law states that

$$PV = RT, \tag{12}$$

where R is a constant independent of the state of the gas. The kinetic energy U of the gas in the cylinder is given by an equation which we borrow from physics:

$$U = C_v\,T,$$

where C_v is a constant that depends upon the nature of the gas but not upon its state. Let us explain the physical significance of the constant C_v. Imagine that we warm the gas from a temperature T_1 to a temperature T_2 while keeping the volume fixed, that is, without moving the piston. Then the kinetic energy of the gas is increased by the amount $C_v(T_2 - T_1)$. In the process of this heating, the gas does no work (the piston does not move). Thus all of the heat given to the gas goes into raising its temperature. That is, C_v is the heat capacity of our gas for fixed volume.

Let σ be the area of a cross section of the cylinder and let l be the distance from the base of the cylinder to the piston. Thus we have $V = \sigma l$. Now let us compress the gas a little, by changing the position of the piston (that is, changing l) by a small negative amount δl. The volume V of the gas then changes by the negative quantity δV, where $\delta V = \sigma \delta l$. While this compression is going on, the force exerted on the piston by the pressure of the gas is approximately equal to σP. Actually the pressure changes a little during the compression, so that the force is equal to $F = \sigma(P + \varepsilon)$, where ε goes to zero along with δl. The work expended by the piston as it moves is equal to the force exerted times the distance travelled, $-\delta l$: this is according to a well-known law of physics. Thus the work is equal to

$$-P\sigma(\delta l + o(\delta l)),$$

which is to say

$$-P\,\delta V + o(\delta V).$$

Since there is no friction and no heat exchange, all of this work turns into kinetic energy of the gas. That is, the kinetic energy of the gas, $U = C_v T$, must increase by an amount $C_v\,\delta T$. We find accordingly that

$$C_v\,\delta T = -P\,\delta V + o(\delta V). \tag{13}$$

By the ideal gas law (12), we have

$$P = \frac{RT}{V}.$$

Substitute this value for P in the formula (13) and obtain

$$C_v\,\delta T = -RT\frac{\delta V}{V} + o(\delta V). \tag{14}$$

In the process of compression of the gas, the temperature T is a function of the volume of the gas. Let us divide (14) by $T\delta V$ to obtain

$$\frac{C_v}{T} \cdot \frac{\delta T}{\delta V} = -\frac{R}{V} + \frac{o(\delta T)}{T\delta V}.$$

In this expression, form the limit as $\delta V \to 0$. This gives us

$$\frac{C_v}{T}\frac{dT}{dV} = -\frac{R}{V},$$

or

$$\frac{C_v}{T}\frac{dT}{dV} + \frac{R}{V} = 0. \tag{15}$$

The main result of the preceding section, §23, shows that the indefinite integral of the left side of (15) is a constant:

$$\int \left(\frac{C_v}{T}\frac{dT}{dV} + \frac{R}{V} \right) dV = c,$$

or

$$\int \frac{C_v\,dT}{T} + \int \frac{R\,dV}{V} = c.$$

Formula (10) gives

$$C_v \ln T + R \ln V = c.$$

Divide this identity by C_v and raise e to the powers of both sides. We get

$$TV^{R/C_v} = \hat{c},$$

where $\hat{c}$ is a constant arising from integration and thus dependent upon the nature of the gas in our cylinder, but independent of its state. Replace T in the last equality by its value from (12), and find

$$PV^{(R+C_v)/C_v} = \tilde{c}, \tag{16}$$

where $\tilde{c}$ again is a constant depending only upon the nature of the gas in the cylinder.

The quantity C_v, as we have already remarked, is the heat capacity of the gas for constant volume. The quantity $C_p = R + C_v$ is the heat capacity of the gas for constant pressure. For, suppose that we heat the gas while keeping constant pressure. The heat expended goes partly into raising the kinetic energy of the gas (that is, raising its temperature) and partly turns into mechanical work, pushing the piston.

We can thus write (16) in the form

$$PV^\gamma = \tilde{c}, \tag{17}$$

where $\gamma = \dfrac{C_p}{C_v}$.

Formula (17) is called *the formula for adiabatic compression of a gas.*

Example 2. Let $u(z)$ be a polynomial of degree n in the variable z. We will use the formula for integration by parts to find the integral of the function $f(z) = u(z) e^z$. To do this, we set $v(z) = e^z$, so that $dv(z) = e^z dz$. The formula for integration by parts yields

$$\int f(z)\,dz = \int u(z)\,e^z\,dz = \int u(z)\,dv(z) = c + u(z)\,e^z - \int e^z\,du(z)$$
$$= c + u(z)\,e^z - \int u'(z)\,e^z\,dz.$$

Therefore, we have reduced our problem to the problem of integrating $u'(z)\,e^z$. Since $u(z)$ is a polynomial, its derivative $u'(z)$ is a polynomial of degree one less than the degree of $u(z)$. We can therefore apply the formula for integration by parts over and over, until we reach the constant polynomial $u^{(n)}(z)$ (by this we mean the nth derivative of $u(z)$). Our final result is

$$\int u(z)\,e^z\,dz = c + u(z)\,e^z - u'(z)\,e^z + u''(z)\,e^z - \cdots + (-1)^n\,u^{(n)}(z)\,e^z$$
$$= c + e^z(u(z) - u'(z) + u''(z) - \cdots + (-1)^n\,u^{(n)}\,z)).$$

The higher derivatives $u'''(z)$ and so on that we use in the above formula are defined in an obvious way. We give more details in §26, (2).

Example 3. By using the formula (1) for integrating a power series term by term, we can obtain power series expansions for the functions $\ln(1 + z)$ and $\arctan z$. By (27) and (29) in §22, we have

$$\ln'(1 + z) = \frac{1}{1 + z} \quad \text{and} \quad \arctan' z = \frac{1}{1 + z^2}.$$

Formula (21) of §17 shows that the rational functions appearing in the last identities have power series expansions with radius 1:

$$\frac{1}{1 + z} = 1 - z + z^2 - \cdots + (-1)^n\,z^n + \cdots$$

and

$$\frac{1}{1 + z^2} = 1 - z^2 + z^4 - \cdots + (-1)^n\,z^{2n} + \cdots.$$

We integrate these power series term by term as in (1) to obtain power series expansions for the functions $\ln(1 + z)$ and $\arctan z$ having radii of convergence equal to 1. To complete the analysis, we need only determine the constants c that appear on the right side of (1), in both cases. To find the values of c, we need only choose some fixed value of z ($z = 0$ is the most convenient) and compute the values of both the left and right sides for this z. Thus the constant c is $\ln(1)$ in the first case and $\arctan 0$ in the second. Both of the functions $\ln z$ and $\arctan z$ are multiple-valued. Thus we find

$\ln 1 = 2k\pi i$ and $\arctan 0 = k\pi$, where k can be an arbitrary integer. Take any of these values in both cases, and get the appropriate value of c to choose in formula (1), since the rest of the power series has value 0 for $z=0$.

For example, we may take the branch of the logarithm with the value 0 at 1 and the branch of the arctangent with value 0 at 0. We get the following power series expansions for these functions in the interior of the unit disk:

$$\ln(1+z) = z - \frac{z^2}{2} + \frac{z^3}{3} - \cdots + (-1)^n \frac{z^{n+1}}{n+1} + \cdots;$$

$$\arctan z = z - \frac{z^3}{3} + \frac{z^5}{5} - \cdots + (-1)^n \frac{z^{2n+1}}{2n+1} + \cdots.$$

§25. The Definite Integral

In §23, we introduced the operation of integration as the inverse of the operation of differentiation. Given a function $f(z)$, we defined its integral as a function $\chi(z)$ with the property that

$$\chi'(z) = f(z). \tag{1}$$

Recall that any function $\chi(z)$ satisfying (1) is called *a primitive of the function $f(z)$.* Formula (1) is in fact an equation in which $f(z)$ is a known function and $\chi(z)$ is the unknown function, which is to be found.

We saw at once that the equation (1) does not have a unique solution. For, if a function $\chi(z)$ is a solution of (1), then so are all functions $\chi(z)+c$, where c is an arbitrary constant. We determined all of the solutions of (1), but only with some complicated considerations. Namely, if the set of numbers (real or complex) where $f(z)$ is defined is a connected set, then every solution of (1) is obtained from any fixed solution of (1) by adding constants. Thus in order to determine all functions $h(z)$ such that

$$h'(z) = f(z), \tag{2}$$

we need to find only one such function, say $\chi(z)$, and then we find that

$$h(z) = \chi(z) + c. \tag{3}$$

In other words, we can determine a "unique" solution of (1) by finding some solution and then choosing the constant c.

There is a different method of determining a fixed solution of (2), which is in some contexts more important. We now take up this method.

Limits of Integration. Let $f(z)$ be a real- or complex-valued function of a real or complex variable. Suppose that $f(z)$ is defined on a connected set in the real number system or the complex number system. Choose any fixed point z_0 where the function $f(z)$ is defined, and select the (unique) indefinite integral $h(z)$ of the function $f(z)$ (compare (2) and (3)) for which

$$h(z_0) = 0. \tag{4}$$

Such an integral exists, if $f(z)$ has any integrals at all. For, given an indefinite integral $\chi(z)$, determine the constant c by the equation

$$\chi(z_0) + c = 0,$$

and then we consider the integral

$$h(z) = \chi(z) + c = \chi(z) - \chi(z_0). \tag{5}$$

The function $h(z)$ satisfies both (2) and (4). We call the function $h(z)$ a *definite integral of $f(z)$* and write it as

$$h(z) = \int_{z_0}^{z} f(\zeta)\, d\zeta. \tag{6}$$

We call z_0 *the lower limit of integration* and z *the upper limit of integration*. The function $h(z)$ which we designate by the expression (6) depends upon the integrand $f(z)$, the lower limit of integration z_0, and the upper limit of integration z. It does not depend in any way on the symbol ζ, which is frequently called *the variable of integration* or *a dummy variable*. We could use any other symbol or letter instead of ζ: for example, the expression

$$\int_{z_0}^{z} f(\zeta)\, d\zeta = \int_{z_0}^{z} f(t)\, dt$$

is a perfectly valid identity.

It is easy to see that the definite integral is completely determined by conditions (2) and (4). We get the same function if we start with another primitive of $f(z)$ than the primitive $\chi(z)$ that we happened to choose.

The Simplest Properties of the Definite Integral. Let $f_1(z)$ and $f_2(z)$ be two functions defined on the same connected set. Let $\chi_1(z)$ and $\chi_2(z)$ be primitives of $f_1(z)$ and $f_2(z)$ respectively, and let α_1 and α_2 be any two numbers. The function $\alpha_1 \chi_1(z) + \alpha_2 \chi_2(z)$ is a primitive of the function $\alpha_1 f_1(z) + \alpha_2 f_2(z)$. It follows that

$$\int_{z_0}^{z} [\alpha_1 f_1(\zeta) + \alpha_2 f_2(\zeta)]\, d\zeta = [\alpha_1 \chi_1(z) + \alpha_2 \chi_2(z)] - [\alpha_1 \chi_1(z_0) + \alpha_2 \chi_2(z_0)]$$

$$= \alpha_1 [\chi_1(z) - \chi_1(z_0)] + \alpha_2 [\chi_2(z) - \chi_2(z_0)]$$

$$= \alpha_1 \int_{z_0}^{z} f_1(\zeta)\, d\zeta + \alpha_2 \int_{z_0}^{z} f_2(\zeta)\, d\zeta.$$

We obtain finally

$$\int\limits_{z_0}^{z} [\alpha_1 f_1(\zeta) + \alpha_2 f_2(\zeta)]\, d\zeta = \alpha_1 \int\limits_{z_0}^{z} f_1(\zeta)\, d\zeta + \alpha_2 \int\limits_{z_0}^{z} f_2(\zeta)\, d\zeta.$$

The lower and upper bounds of integration, z_0 and z, are to a certain extent of equal status. To emphasize this, we may write the upper limit of integration as z_1:

$$\int\limits_{z_0}^{z_1} f(\zeta)\, d\zeta = \chi(z_1) - \chi(z_0). \qquad (7)$$

This is simply the identity (5) rewritten. We may interchange the limits of integration and obtain

$$\int\limits_{z_1}^{z_0} f(\zeta)\, d\zeta = \chi(z_0) - \chi(z_1). \qquad (8)$$

From (7) and (8) we find that

$$\int\limits_{z_1}^{z_0} f(\zeta)\, d\zeta = - \int\limits_{z_0}^{z_1} f(\zeta)\, d\zeta. \qquad (9)$$

Let z_0, z_1, and z_2 be three points where $f(z)$ is defined. We have

$$\int\limits_{z_0}^{z_1} f(\zeta)\, d\zeta = \chi(z_1) - \chi(z_0),$$

$$\int\limits_{z_1}^{z_2} f(\zeta)\, d\zeta = \chi(z_2) - \chi(z_1),$$

$$\int\limits_{z_0}^{z_2} f(\zeta)\, d\zeta = \chi(z_2) - \chi(z_0).$$

These three equalities yield the important fact that

$$\int\limits_{z_0}^{z_1} f(\zeta)\, d\zeta + \int\limits_{z_1}^{z_2} f(\zeta)\, d\zeta = \int\limits_{z_0}^{z_2} f(\zeta)\, d\zeta. \qquad (10)$$

Uniqueness of the Definite Integral. We suppose as before that the original function $f(z)$ that we are integrating is defined on a connected set. We write

$$\int\limits_{z_0}^{z_1} f(\zeta)\, d\zeta = h(z_0, z_1). \qquad (11)$$

As in (5), we have

$$h(z_0, z_1) = \chi(z_1) - \chi(z_0).$$

The properties (2) and (4) may be written as

$$\frac{\partial h(z_0, z_1)}{\partial z_1} = f(z_1) \tag{12}$$

and

$$h(z_0, z_0) = 0, \tag{13}$$

respectively. Since (2) and (4) determine $h(z)$ uniquely, the relations (12) and (13) determine $h(z_0, z_1)$ uniquely.

We may interchange z_0 and z_1 in (12) and (13):

$$\frac{\partial h(z_1, z_0)}{\partial z_0} = f(z_0) \tag{14}$$

and

$$h(z_1, z_1) = 0. \tag{15}$$

These relations determine the function $h(z_1, z_0)$ uniquely. We now observe that

$$h(z_1, z_0) = -h(z_0, z_1) \tag{16}$$

(see (9) and (11)). That is to say, the functions $h(z_0, z_1)$ and $h(z_1, z_0)$ determine each other. In the relations (14) and (15), we replace $h(z_1, z_0)$ using (16). We find that

$$\frac{\partial h(z_0, z_1)}{\partial z_0} = -f(z_0) \tag{17}$$

and

$$h(z_1, z_1) = 0. \tag{18}$$

The properties (17) and (18) determine $h(z_0, z_1)$ uniquely. Summarizing, we may say that the function $h(z_0, z_1)$ satisfies the pair of relations (12) and (13) and also the pair of relations (17) and (18). Each pair determines the function uniquely.

The Existence of the Primitive. Everything done so far in this section rests upon the hypothesis that the function $f(z)$ admits some primitive $\chi(z)$. If the function $h(z_0, z_1)$ satisfies (12), then it is a primitive of the function $f(z_1)$. Similarly, if $h(z_0, z_1)$ satisfies (17), then $-h(z_0, z_1)$ is a primitive of the function $f(z_0)$. Summarizing, we may say that the definite integral

$$\int_{z_0}^{z_1} f(\zeta) \, d\zeta$$

has been found if we can find a function $h(z_0, z_1)$ that satisfies either (12) and (13) or (17) and (18).

Change of the Variable of Integration. We have already studied change of variable in computing indefinite integrals (see §24, (8)). In carrying this technique over to definite integrals, we must also pay attention to what happens to the limits of integration. Consider a definite integral

$$h(z) = \int_{z_0}^{z} f(\zeta)\, d\zeta.$$

Instead of z, we introduce a new variable w by the formula

$$z = \psi(w),$$

where $\psi(w)$ is an invertible function, that is, there is the equivalent formula

$$w = \varphi(z),$$

so that the identities

$$\varphi(\psi(w)) = w \quad \text{and} \quad \psi(\varphi(z)) = z \tag{19}$$

obtain. We will explain how the function

$$h^*(w) = h(\psi(w))$$

can be written as a definite integral. We will show that

$$h^*(w) = \int_{w_0}^{w} g(\omega)\, d\omega, \tag{20}$$

where

$$\left.\begin{array}{l} g(\omega) = f(\psi(\omega))\,\psi'(\omega), \\[2mm] w_0 = \varphi(z_0), \quad w = \varphi(z). \end{array}\right\} \tag{21}$$

and

To prove (20), we will show that the function $h^*(w)$ satisfies conditions (12) and (13). We use the formula for differentiating a composite function (§22, (21)) to write

$$(h^*(w))' = (h(\psi(w)))' = h'(\psi(w)) \cdot \psi'(w) = f(\psi(w)) \cdot \psi'(w) = g(w):$$

compare this with (21). This is condition (12) for the function $h^*(w)$. We also have

$$h^*(w_0) = h(\psi(w_0)) = h(z_0) = 0.$$

Thus condition (13) holds as well. We have thus proved that $h^*(w)$ is the definite integral (20).

Dependence on a Parameter. Suppose that the integrand depends not only on the variable z with respect to which we integrate but also upon another variable u, which we will call *a parameter*. Thus the function f has the form $f(z, u)$. The primitives of $f(z, u)$ also depend upon u, having the form $\chi(z, u)$. Thus we have

$$\frac{\partial}{\partial z} \chi(z, u) = f(z, u).$$

All other functions $h(z, u)$ such that

$$\frac{\partial}{\partial z} h(z, u) = f(z, u) \tag{22}$$

differ from the function $\chi(z, u)$ by an additive quantity that is independent of z but which may well depend upon u. Thus the function $h(z, u)$ has the form

$$h(z, u) = \chi(z, u) + c(u),$$

where $c(u)$ is an arbitrary function of the parameter u. Choose an arbitrary but fixed value z_0 where $f(z, u)$ is defined, and require that the function $h(z, u)$ satisfy the condition

$$h(z_0, u) = 0. \tag{23}$$

For this to happen we must choose the function $c(u)$ of the parameter u in such a way that

$$\chi(z_0, u) + c(u) = 0,$$

that is,

$$c(u) = -\chi(z_0, u).$$

The function $h(z, u)$ that satisfies both (22) and (23) accordingly has the form

$$h(z, u) = \chi(z, u) - \chi(z_0, u). \tag{24}$$

The function $h(z, u)$ is a definite integral of the function $f(z, u)$ and is written as

$$h(z, u) = \int_{z_0}^{z} f(\zeta, u) \, d\zeta. \tag{25}$$

Thus the definite integral (25) has the form (24). To underline the equal status of the lower and upper limits of integration, we write the upper limit z as z_1:

$$\int_{z_0}^{z_1} f(\zeta, u) \, d\zeta = h(z_0, z_1, u) = \chi(z_1, u) - \chi(z_0, u). \tag{26}$$

This formula is designed to emphasize the dependence of the integral (26) upon z_0, z_1, and u. As a function of the upper limit of integration, the definite integral $h(z_0, z_1, u)$ satisfies the conditions

$$\frac{\partial}{\partial z_1} h(z_0, z_1, u) = f(z_1, u) \tag{27}$$

and

$$h(z_0, z_0, u) = 0. \tag{28}$$

These conditions define the definite integral uniquely. Considered as a function of the lower limit of integration, the definite integral $h(z_0, z_1, u)$ satisfies the conditions

$$\frac{\partial}{\partial z_0} h(z_0, z_1, u) = -f(z_0, u) \tag{29}$$

and

$$h(z_1, z_1, u) = 0, \tag{30}$$

which again define it uniquely. Thus, to prove that a certain function $h(z_0, z_1, u)$ of three variables is actually a definite integral,

$$h(z_0, z_1, u) = \int_{z_0}^{z_1} f(\zeta, u) \, d\zeta,$$

it suffices to show that it satisfies either the pair of conditions (27) and (28) or the pair of conditions (29) and (30). We must simply make a lucky guess to find the function $h(z_0, z_1, u)$ that satisfies one of the two pairs of conditions.

All of what we have said about integrals of functions depending on a parameter is of little interest by itself. The situation becomes nontrivial and demands particular attention, however, if the parameter u coincides with one of the limits of integration:

$$\text{either } u = z_0 \quad \text{ or } u = z_1.$$

In this case, we are required to find one of the integrals

$$h(z_0, z_1, z_1) = \int_{z_0}^{z_1} f(\zeta, z_1) \, d\zeta \tag{31}$$

or

$$h(z_0, z_1, z_0) = \int_{z_0}^{z_1} f(\zeta, z_0) \, d\zeta. \tag{32}$$

In finding the integral (31), checking conditions (27) and (28) will not work, but checking conditions (29) and (30) will. In finding the integral (32), on the other hand, we must use (27) and (28). Let us explain why this is so, considering the case of the first integral, (31). In the function $h(z_0, z_1, u)$, we have replaced u by z_1, and the function $h(z_0, z_1, z_1)$ considered as the integral (31) simply does not satisfy the condition

$$\frac{\partial}{\partial z_1} h(z_0, z_1, z_1) = f(z_0, z_1). \tag{33}$$

In obtaining (27), we held u constant, and in the left side of (33) we must differentiate with respect to $u = z_1$. Conditions (29) and (30), on the other hand, are perfectly reasonable for the integral (31):

$$\frac{\partial}{\partial z_0} h(z_0, z_1, z_1) = -f(z_0, z_1) \tag{34}$$

and

$$h(z_1, z_1, z_1) = 0. \tag{35}$$

The problem is analogous for finding the integral (32). We must show that

$$\frac{\partial}{\partial z_1} h(z_0, z_1, z_0) = f(z_0, z_1)$$

and that

$$h(z_0, z_0, z_0) = 0.$$

Integrating Inequalities. In this paragraph, we will consider only real-valued functions of a real variable. We will show that if two functions $f_1(x, u)$ and $f_2(x, u)$ satisfy the inequality

$$f_1(x, u) \le f_2(x, u) \tag{36}$$

for all relevant values of x and u, then the inequality

$$\mathrm{sign}(x - x_0) \int_{x_0}^{x} f_1(\xi, u) \, d\xi \le \mathrm{sign}(x - x_0) \int_{x_0}^{x} f_2(\xi, u) \, d\xi \tag{37}$$

holds for all relevant values of u. That is, an inequality of the form (36) can be integrated between the limits x_0 and x.

To prove (37), we consider the function

$$f(x, u) = f_2(x, u) - f_1(x, u).$$

To prove (37) it suffices to show that if

$$f(x, u) \ge 0 \tag{38}$$

for all relevant values of x and u, then

$$\operatorname{sign}(x-x_0)\int_{x_0}^{x} f(\xi,u)\,d\xi \geq 0 \tag{39}$$

for all relevant values of u. We first write

$$h(x,u)=\int_{x_0}^{x} f(\xi,u)\,d\xi.$$

From (23) we see that

$$h(x,u)=h(x,u)-h(x_0,u). \tag{40}$$

In view of Lagrange's formula (see §23, (20)), we have

$$h(x,u)-h(x_0,u)=\frac{\partial}{\partial x}h(\eta,u)(x-x_0)=f(\eta,u)(x-x_0) \tag{41}$$

for some number η such that $x_0<\eta<x$. Since $f(\eta,u)\geq 0$, in view of (38), (40) and (41) show that

$$\operatorname{sign}(x-x_0)\cdot h(x,u)\geq 0:$$

and this is the required inequality (39).

Note that (39) and hence (37) hold in the special case $u=x_0$. We can use the conditions (27) and (28), which identify the definite integral. Using (29) and (30) instead, we can prove (37) if we set $u=x$.

Integrating the Modulus of a Function. Suppose that

$$|f(\xi,u)|\leq M \tag{42}$$

for all relevant values of u and all ξ in the closed interval $[x_0 x]$. Then we have

$$\left|\int_{x_0}^{x} f(\xi,u)\,d\xi\right|\leq M\cdot|x-x_0|. \tag{43}$$

To prove this, rewrite (42) as

$$-M\leq f(\xi,u)\leq M.$$

Integrate these two inequalities between the limits x_0 and x and apply the result of the preceding paragraph. We get the two inequalities

$$\operatorname{sign}(x-x_0)\cdot(-M)(x-x_0)\leq\operatorname{sign}(x-x_0)\int_{x_0}^{x} f(\xi,u)\,d\xi$$

$$\leq\operatorname{sign}(x-x_0)\cdot M\cdot(x-x_0). \tag{44}$$

From these inequalities, (42) is immediate.

§26. Taylor Series

Expansion of a function in a power series is one of the most important ways in which we can compute it for various values of its argument. Sometimes we are so to say given the power series expansion of a function. Thus we actually defined the exponential function $\exp(z)$ by its power series expansion (see §18). Sometimes a function may be given to us in some other form, and by some maneuver we are able to expand it in a power series, from which we can compute its values. This is the case with the functions $\cos x$ and $\sin x$ of a real variable x. These functions are known from elementary trigonometry, being defined geometrically. We obtained their power series expansions in §19, by making use of properties of the function $\exp(z)$. Using the expansion of a function in a power series, we can often deduce its properties, as we did in §18 with the function $\exp(z)$. We listed a number of its properties at the very beginning of §18. If we have a power series expansion of a function for real values of its argument, we can immediately extend this function to an entire open disk in the complex plane, as we showed in §20. We have already found power series expansions for a number of transcendental functions. It is now time for us to take up the power series expansion of an arbitrary function. We will do this in the present section. For the sake of maximum generality, we will consider expansions in powers of $z - z_0$, where z_0 is some fixed point where the function is defined. We will deal with the real and complex cases simultaneously: all of our functions may be real- or complex-valued, and they may be defined only for real values or for complex values.

We will suppose at first that we have a function $f(z)$ that can be expanded in a power series in the variable $z - z_0$:

$$f(z) = a_0 + a_1(z - z_0) + a_2(z - z_0)^2 + \cdots + a_n(z - z_0)^n + \cdots. \tag{1}$$

This series of course converges for $z = z_0$. For the equality (1) to make any sense, the series must converge for some values of z different from z_0, which is to say that the series (1) has a positive radius of convergence. Thus there is a number $r > 0$ (which may be $+\infty$) such that the series (1) converges absolutely for all z such that

$$|z - z_0| < r.$$

The coefficients of the series (1) can be easily computed from the derivatives of the function $f(z)$. We need not only the first derivative of $f(z)$, but the derivatives of all orders. Since previously we have touched only cursorily on this topic, let us give a formal definition.

Derivatives of Different Orders. We already know what the derivative $f'(z)$ of the function $f(z)$ means. We may call it *the first derivative of* $f(z)$.

The derivative $(f'(z))'$ of the first derivative is called *the second derivative of* $f(z)$. We write it as $f''(z)$. Similarly the derivative $(f''(z))'$ is called *the third derivative of* $f(z)$. It is denoted by $f'''(z)$ or $f^{(3)}(z)$. The nth derivative of $f(z)$ is defined inductively by the equality

$$f^{(n)}(z) = (f^{(n-1)}(z))'. \tag{2}$$

We already know that if a power series converges in an open disk of positive radius r, then the derived series, obtained by differentiating the series term by term, also converges in the same open disk and is equal at every point of the open disk to the derivative $f'(z)$. See §22. We did this, it is true, for functions (1) with $z_0 = 0$. We can do exactly the same for functions (1) with arbitrary z_0, in view of (22) in §22, since

$$\frac{d(z - z_0)}{dz} = 1.$$

The nth derivative of the function (1) is accordingly

$$f^{(n)}(z) = n!\, a_n + (n+1)n(n-1)\ldots 2a_{n+1}(z - z_0)$$
$$+ \cdots + (n+k)(n+k-1)\ldots(k+1)a_{n+k}(z - z_0)^k + \cdots.$$

This identity is valid for all z such that $|z - z_0| < r$. Set $z = z_0$ and find that $f^{(n)}(z_0) = n!\, a_n$, so that

$$a_n = \frac{f^{(n)}(z_0)}{n!}. \tag{3}$$

That is, the coefficients of the power series (1) can be computed from the successive derivatives of $f(z)$, computed at z_0. We may rewrite (1) as

$$f(z) = f(z_0) + \frac{f'(z_0)}{1!}(z - z_0) + \cdots + \frac{f^{(n)}(z_0)}{n!}(z - z_0)^n + \cdots. \tag{4}$$

The series on the left side of (4) is called *the Taylor series for the function* $f(z)$ *at the point* $z = z_0$. Plainly the Taylor series expansion of $f(z)$ at $z = z_0$ can exist only if $f(z)$ admits derivatives of all orders at z_0. Actually we assumed more than this in establishing (4), *viz.*, that $f(z)$ admits a power series expansion (1) with positive radius of convergence.

The question now arises: can a given function $f(z)$ be expanded in a Taylor series at the point $z = z_0$? We must study the difference

$$R_n = R_n(z_0, z)$$
$$= f(z) - \left[f(z_0) + \frac{1}{1!}f'(z_0)(z - z_0) + \cdots + \frac{1}{n!}f^{(n)}(z_0)(z - z_0)^n \right]. \tag{5}$$

This of course is the difference between $f(z)$ and the nth partial sum of the series appearing in (4). Clearly the series expansion (4) holds if and only if

$$\lim_{n \to \infty} R_n(z_0, z) = 0. \tag{6}$$

We will find that the quantity $R_n(z_0, z)$, often called *the remainder term for* (4), can be written as the definite integral of a comparatively simple function. This representation of the remainder allows us in a number of cases to settle the question of whether or not the expansion (4) obtains. To write $R_n(z_0, z)$ as a definite integral, we will use the conditions (34) and (35) of §25, setting

$$R_n(z_0, z) = h(z_0, z, z).$$

To verify condition (34) of §25, we must compute the partial derivative

$$\frac{\partial}{\partial z_0} R_n(z_0, z). \tag{7}$$

This will enable us to find the function $f(z_0, z)$ on the right side of (34). Then we must verify condition (35).

To compute the partial derivative (7), we of course regard z_0 as variable and z as fixed. We find first

$$\frac{\partial}{\partial z_0} f(z) = 0, \qquad \frac{\partial}{\partial z_0} f(z_0) = f'(z_0). \tag{8}$$

For $k > 0$, we get

$$\frac{\partial}{\partial z_0} \frac{f^{(k)}(z_0)}{k!} (z - z_0)^k = -\frac{1}{(k-1)!} f^{(k)}(z_0)(z - z_0)^{k-1}$$

$$+ \frac{1}{k!} f^{(k+1)}(z_0)(z - z_0)^k. \tag{9}$$

Now sum the equalities (9) from 1 to n and take account of (8). We find that

$$\frac{\partial}{\partial z_0} R_n(z_0, z) = -\frac{1}{n!} f^{(n+1)}(z_0)(z - z_0)^n.$$

To verify condition (35) of §25, we must replace z_0 by z in (5). We obviously obtain

$$R_n(z, z) = 0.$$

The results of §25 now make it clear that

$$R_n = R_n(z_0, z) = \frac{1}{n!} \int_{z_0}^{z} f^{(n+1)}(\zeta)(z - \zeta)^n \, d\zeta. \tag{10}$$

Combining (5) and (10), we obtain the equality

$$f(z)=f(z_0)+\frac{1}{1!}f'(z_0)(z-z_0)+\cdots+\frac{1}{n!}f^{(n)}(z_0)(z-z_0)^n$$

$$+\frac{1}{n!}\int_{z_0}^{z}f^{(n+1)}(\zeta)(z-\zeta)^n\,d\zeta. \tag{11}$$

The identity (11) is called *Taylor's formula*. It is valid when $f(z)$ has all of its derivatives up to and including the derivative $f^{(n+1)}(z)$. The equality (11) is important not only for establishing expansibility in power series. In order for the equality (6) to make sense, $f(z)$ must have derivatives of all orders. Thus the function $f(z)$ admits a Taylor series expansion if and only if it has derivatives of all orders and the series (4) converges, or equivalently, (6) holds. It does not suffice, however, to demand merely that the series (4) converge. See Example 2 below, where we construct a function with derivatives of all orders for which the series (4) converges for all z (we take z_0 $=0$), and yet the function is not expansible in a Taylor series in the neighborhood of $z_0=0$.

We will use the results just obtained to find Taylor series expansions for the real-valued function

$$f(x)=(1+x)^r, \tag{12}$$

where x is a real variable and r is an arbitrary real constant (see §22, (30) and (31)). The Taylor series expansion (about 0) is:

$$(1+x)^r=1+\frac{r}{1!}x+\frac{r(r-1)}{2!}x^2+\cdots+\frac{r(r-1)\ldots(r-n+1)}{n!}x^n+\cdots. \tag{13}$$

When r is a natural number, (13) coincides with Newton's binomial formula. It is natural to preserve this designation for the (usually infinite) series expansion (13), especially as it was known to Newton. We will prove that the expansion (13) holds for all real x such that $|x|<1$. We first compute the nth derivative of the function $(1+x)^r$, using formula (31) of §22. We find

$$f^{(n)}(x)=r(r-1)\cdots(r-n+1)(1+x)^{r-n}.$$

We wish to prove that the remainder term goes to zero as $n\to\infty$, for all x such that $|x|<1$. To do this, we will estimate the quantity

$$\mu=\frac{x-\xi}{1+\xi}, \tag{14}$$

where ξ is a real variable assuming values in the closed interval $[0x]$. We will prove that

$$|\mu|\leq|x|. \tag{15}$$

If x is positive, then ξ is positive, and the estimate (15) is immediate from the definition (14). If x is negative, then ξ is also negative, and (15) follows from the identity

$$\mu = x \cdot \frac{1 - \dfrac{\xi}{x}}{1 + \xi}.$$

For our function (12), the remainder term R_n is given by the formula

$$R_n = A_n \int_0^x (1+\xi)^{r-n-1}(x-\xi)^n \, d\xi = A_n \int_0^x \mu^n (1+\xi)^{r-1} \, d\xi,$$

where

$$A_n = \frac{r(r-1)\cdots(r-n+1)}{n!}.$$

Since $|x| < 1$ and the variable ξ runs between 0 and x, the quantity $(1+\xi)^{r-1}$ remains less than a certain constant δ. Our estimates for the integral (§ 25, (43)) show that

$$|R_n| \leq |A_n| \, \delta |x|^{n+1}.$$

Let us write A_n as

$$A_n = r \cdot \frac{r-1}{1} \cdot \frac{r-2}{2} \cdot \ldots \cdot \frac{r-n}{n}.$$

Thus, except for the factor r, A_n is the product of numbers of the form

$$\frac{r-k}{k} = \frac{r}{k} - 1.$$

It is clear that

$$\lim_{k \to \infty} \frac{r}{k} - 1 = -1. \tag{16}$$

We now choose a positive number s such that

$$|x| < s < 1.$$

From (16) we see that there exists a positive integer l such that

$$\text{if } k > l, \text{ we have } \left| \frac{r-k}{k} x \right| < s.$$

Putting all this together, we find

$$|R_n| \leqq \left| \frac{r(r-1)\cdots(r-l)}{l!} \right| \cdot \delta \cdot s^{n-l} x^{l+1}.$$

Since $s < 1$, we find that $s^{n-l} \to 0$ as $n \to \infty$. This proves (6) for our function $f(x)$. That is, (13) holds for $|x| < 1$.

Example 1. Let us now write R_n, the remainder term in Taylor's formula, in a simple and frequently used form, for the case in which $f(x)$ is a real-valued function of a real variable. We will suppose that the function $f^{(n+1)}(\xi)$ is continuous on the closed interval from x_0 to x and that its minima and maxima respectively are p and q:

$$p \leq f^{(n+1)}(\xi) \leq q. \tag{17}$$

We will prove that

$$R_n = \frac{(x-x_0)^{n+1}}{(n+1)!} \gamma, \tag{18}$$

where γ is a number such that $p \leq \gamma \leq q$. We can also write the number γ in the form

$$\gamma = f^{(n+1)}(\eta),$$

where η is a number between x_0 and x. (For a proof that such an η exists, see Example 3 below.) Then R_n assumes the commonly written form

$$R_n = \frac{(x-x_0)^{n+1}}{(n+1)!} f^{(n+1)}(\eta). \tag{19}$$

To prove (19) we will use the identity

$$\int_{x_0}^{x} (x-\xi)^n \, d\xi = \frac{1}{(n+1)} (x-x_0)^{n+1}.$$

Knowing that the signs of $(x-\xi)$ and $(x-x_0)$ are the same, we multiply the two inequalities (17) by the positive number $(x-\xi)^n \operatorname{sign}(x-x_0)^n$, obtaining

$$(x-\xi)^n p \operatorname{sign}(z-x_0)^n \leq (x-\xi)^n f^{(n+1)}(\xi) \operatorname{sign}(x-x_0)^n$$
$$\leq (x-\xi)^n q \operatorname{sign}(x-x_0)^n.$$

We integrate this repeated inequality between the limits x_0 and x to obtain, as in §25, (44), the inequalities

$$\frac{p}{n+1}(x-x_0)^{n+1} \operatorname{sign}(x-x_0)^{n+1} \leq \operatorname{sign}(x-x_0)^{n+1} \int_{x_0}^{x} (x-\xi)^n f^{(n+1)}(\xi) \, d\xi$$
$$\leq \frac{q}{n+1}(x-x_0)^{n+1} \operatorname{sign}(x-x_0)^{n+1}.$$

Divide this repeated inequality by $n!$ and obtain

$$\frac{p}{(n+1)!}(x-x_0)^{n+1} \operatorname{sign}(x-x_0)^{n+1} \leq R_n \operatorname{sign}(x-x_0)^{n+1}$$
$$\leq \frac{q}{(n+1)!}(x-x_0)^{n+1} \operatorname{sign}(x-x_0)^{n+1}.$$

The equality (18) follows at once. If we use the remainder R_n in the form (19), it is easy to obtain the Taylor series expansions for e^x, $\sin x$, and $\cos x$ (although of course e^x was originally *defined* as a power series, which must be its Taylor series). The power series expansion (13) for the function $(1+x)^r$, on the other hand, cannot be obtained in any easy way from (19).

Example 2. We give here an example of a function that has derivatives of all orders everywhere and yet is at one point not expansible in a Taylor series. We begin with the observation that

$$\lim_{\xi \to \infty} \frac{e^{\xi}}{\xi^n} = \infty$$

for all positive integers n. This follows at once from the estimates

$$\frac{e^{\xi}}{\xi^n} = \xi^{-n}\left(1 + \frac{\xi}{1!} + \frac{\xi^2}{2!} + \cdots + \frac{\xi^n}{n!} + \cdots\right) > \frac{\xi}{(n+1)!}$$

for all positive ξ. Taking reciprocals, we see that $e^{-\xi}$ goes to zero faster than $\frac{1}{\xi^n}$ as $\xi \to \infty$. Now set $\xi = \frac{1}{x^2}$ and infer that for all polynomials P in the variable $\frac{1}{x}$, we have

$$\lim_{x \to 0} P\left(\frac{1}{x}\right) e^{-\frac{1}{x^2}} = 0. \tag{20}$$

We now consider the function $f(x) = e^{-\frac{1}{x^2}}$, defined for all real x except for $x = 0$. We define $f(0)$ to be 0: then $f(x)$ is continuous everywhere, including the point 0. For every nonzero value of x, we compute the derivatives of $f(x)$ by the rules given in §22. Each of them has the form

$$f^{(n)}(x) = P_n\left(\frac{1}{x}\right) e^{-\frac{1}{x^2}},$$

where the exact form of the polynomial $P_n\left(\frac{1}{x}\right)$ need not concern us. From (20), we infer that

$$\lim_{x \to 0} f^{(n)}(x) = 0.$$

We cannot infer from this, however, that $f^{(n)}(0)$ has any prescribed value. We will prove by induction that $f^{(n)}(0)$ exists and is equal to 0 for $n = 1, 2, 3, \ldots$. We proceed by induction. Suppose that $f^{(n-1)}(0) = 0$ (which is true by definition for $n = 1$). We will prove that in this case $f^{(n)}(0) = 0$. We compute as follows:

$$f^{(n)}(0) = \lim_{x \to 0} \frac{f^{(n-1)}(x) - f^{(n-1)}(0)}{x} = \lim_{x \to 0} \frac{1}{x} P_n\left(\frac{1}{x}\right) e^{-\frac{1}{x^2}} = 0$$

(see (20)). Thus the function $f(x)$ and all of its derivatives assume the value 0 at $x=0$. The Taylor series for $f(x)$ with $x_0=0$ is the series identically 0. If $f(x)$ were expansible in a Taylor series at 0, we would have $f(x)\equiv 0$, which is clearly false, as the function e^z never assumes the value 0. Thus $f(x)$ is not expansible in a Taylor series at 0. $\left(\text{Note that } f(x) \text{ is defined only for}\right.$ *real* values of x. The behavior of the function $\exp\left(\dfrac{-1}{z^2}\right)$ near $z=0$ is much more complicated, and there is no way to define this function at 0 so that it is continuous.$\Big)$

Example 3. Consider a real-valued continuous function f of a real variable defined on some connected set of real numbers. Let y_0 and y_2 be two distinct values of f:

$$y_0 = f(x_0) \quad \text{and} \quad y_2 = f(x_2). \tag{21}$$

With no loss of generality we may suppose that $x_0 < x_2$ and that $y_0 < y_2$. (If $y_0 > y_2$, we study the function $-f(x)$.) Then we claim that the function $f(x)$ assumes every value y_1 such that $y_0 < y_1 < y_2$, and indeed does this in the interval $x_0 < x < x_2$. This is called *the intermediate value property for the function $f(x)$*.

We prove this assertion from the existence of suprema for bounded sets of real numbers. Let M be the set of all numbers x such that $x_0 \leq x \leq x_2$ and $f(x) \leq y_1$. The set M is nonvoid, since it evidently contains x_0:

$$f(x_0) = y_0 < y_1.$$

Let x_1 be the supremum of the set M (see §15). We will prove by contradiction that

$$f(x_1) = y_1. \tag{22}$$

We assume that (22) fails, so that

$$\varepsilon = |f(x_1) - y_1| > 0.$$

Since $f(x)$ is continuous, there is a positive number δ such that

$$|f(x) - f(x_1)| < \varepsilon$$

for all x such that $x_0 \leq x \leq x_2$ and

$$|x - x_1| < \delta. \tag{23}$$

Let U denote the set of all points in the closed interval $[x_0 x_2]$ that satisfy (23). If $f(x_1) < y_1$, then all points of the set U belong to M. If, on the other hand, $f(x_1) > y_1$, then all points of the set U are outside of the set M.

Continuity of $f(x)$ shows that $x_0 < x_1 < x_2$ (we leave the details of this simple argument to the reader). In the first case, M contains points to the right of x_1, so that x_1 is not even an upper bound to M. In the second case, there is an interval with right endpoint x_1 that contains no points of M, and so x_1, which is an upper bound for M, is certainly not the least upper bound of M. Thus we have a contradiction, and (22) must hold.

Example 4. We return to Newton's binomial formula in its general form (13). Plainly the series in (13) converges for all complex z such that $|z| < 1$. Thus the radius of convergence of the series (13) is greater than or equal to 1. (It is in fact equal to 1 if r is not a positive integer.) Thus we can define the complex-valued function $(1 + z)^r$ for all complex z such that $|z| < 1$ by

$$(1+z)^r = 1 + \frac{r}{1!}z + \frac{r(r-1)}{2!}z^2 + \cdots + \frac{r(r-1)\cdots(r-n+1)}{n!}z^n + \cdots. \qquad (24)$$

Let us use the expansion (24) to obtain a Taylor series expansion for the function $\arcsin z$. Recall from formula (28) of §22 that

$$\arcsin' z = (1 - z^2)^{-\frac{1}{2}}.$$

Expand the right side of this identity by means of (24):

$$\arcsin' z = 1 + \tfrac{1}{2}z^2 + \cdots + \left(\frac{1}{2}\right)^n \cdot \frac{1 \cdot 3 \cdots (2n-1)}{n!} z^{2n} + \cdots. \qquad (25)$$

Choosing the branch of the function $\arcsin z$ that is zero for $z = 0$, we may integrate the series (25) term by term to find that

$$\arcsin z = z + \frac{1}{2 \cdot 3}z^2 + \cdots + \left(\frac{1}{2}\right)^n \frac{1}{2n+1} \cdot \frac{1 \cdot 3 \cdot \cdots \cdot (2n-1)}{n!} z^{2n+1} + \cdots. \qquad (26)$$

Thus the function $\arcsin z$ admits the Taylor series expansion (26).

Chapter VI
The Integral Calculus

The integral calculus begins with the description of the definite integral of a real-valued function $f(x)$ of a real variable x as the area of the region included between the graph of the function, the axis of abscissas, and two vertical lines. At this point, we raise no question as to the existence of this area, taking it as intuitively obvious. We prove that the definite integral, defined in this fashion, is a primitive for the function $f(x)$. Thus we prove the existence of a primitive, proceeding from the assumption that areas exist. We then describe the definite integral as the limit of a certain sequence of finite sums. This produces a rigorous proof that the definite integral exists. At the end of the chapter, we present some applications of the definite integral, including the calculation of the length of the graph of a function and of a curve given in parametric form.

§27. The Definite Integral as an Area

In §25, we gave a description of the definite integral of a function $f(z)$ in terms of primitives of $f(z)$, as follows. The symbol

$$h(z) = \int_{z_0}^{z} f(\zeta)\, d\zeta$$

denotes the function such that

$$h'(z) = f(z) \quad \text{and} \quad h(z_0) = 0.$$

This description of the definite integral suffers from the defect that it gives no hint of how to construct the definite integral. We must simply make a shrewd guess to find some primitive $\chi(z)$ of the function $f(z)$: $\chi'(z) = f(z)$. This "method" of guessing is by no means bad. Using it, we were able to integrate polynomials, power series, and a large number of elementary transcendental functions (though by no means all). See §24 for the details. But none of this is genuinely constructive: we lack a specific way to find integrals. There is a second method of describing definite integrals for real-valued functions of a real variable in terms of areas. This method was discovered earlier than the first method that we took up. We will now present this "area" definition.

We can define the definite integral as the area of a certain region in the plane, bounded on three sides by straight line segments and on a fourth by a curve. We suppose for the nonce that areas of such regions exist. We then obtain a clear geometric picture of what a definite integral is. When we have given a concrete method of approximating arbitrarily closely to the area of a region in the plane, we will have a method of computing definite integrals arbitrarily closely. We will do this later, in §29. We limit ourselves in this section to a description of the connection between integrals and areas, under the assumption that we know intuitively what areas are.

The Area of a Graph. We start with some known continuous function $f(\xi)$ of a real variable ξ. Let us construct the graph C of this function. We make a sketch as in Fig. 65. In a plane P, we draw the axis of abscissas horizontally and the axis of ordinates vertically. We think of the variable ξ as moving along the axis of abscissas and along the axis of ordinates the quantity

$$y = f(\xi).$$

We denote the point on the axis of abscissas with abscissa equal to ξ by the single symbol ξ (although $(\xi, 0)$ is its accurate designation). Now take any two completely arbitrary values of the argument ξ and denote them by x_0 and x. Draw vertical lines through the points x_0 and x. Let $a_0 = f(x_0)$ and $a = f(x)$ be the points where these two vertical lines meet the graph C. Suppose that the portion $a_0 a$ of the curve C lies entirely on one side of the axis of abscissas. Then we can define a region in the plane P in a wholly natural fashion. This region is bounded on the sides by the line segments $[x_0 a_0]$ and $[xa]$. If C lies above the axis of abscissas, it is bounded above the curve $a_0 a$ and on the bottom by the line segment $[x_0 x]$ on the axis of abscissas. If C lies entirely below the axis of abscissas, then $[x_0 x]$ is the upper boundary of the region and $a_0 a$ is its lower boundary. Although C is in general a curved line, it seems very natural to suppose that the region we

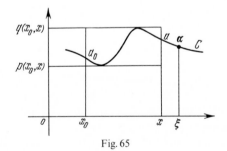

Fig. 65

have just described has an area. This area by its very nature is positive, or at worst zero, as happens when $x = x_0$. We adopt the following conventions regarding the area, so as to make it either positive or negative. Let us denote the region that we have described (it is a sort of generalized quadrilateral) by the symbol $x_0 x a a_0$. For $x_0 < x$, we regard the area of $x_0 x a a_0$ as positive if the curve C lies entirely above the axis of abscissas and as negative if the curve C lies entirely below the axis of abscissas. For $x_0 > x$, we regard the area as negative if C lies entirely above the axis of abscissas and as positive if C lies entirely below the axis of abscissas. The area of $x_0 x a a_0$, with these conventions as to its sign, will be denoted by $h(x_0, x)$.

Let us now compare the area $h(x_0, x)$ with the areas of two rectangles that spring at once to our attention. Let $p(x_0, x)$ be the minimum of the function $f(\xi)$ in the closed interval $[x_0 x]$ and let $q(x_0, x)$ denote its maximum in the same closed interval. Now on the closed interval $[x_0 x]$ as a base, construct a rectangle of height $p(x_0, x)$ and a rectangle of height $q(x_0, x)$. We agree that if the height is positive, then the rectangle lies above the axis of abscissas, while if the height is negative, then the rectangle lies below the axis of abscissas. Suppose that the curve C lies entirely above the axis of abscissas. Then the region $x_0 x a a_0$ contains the first rectangle and is contained within the second rectangle. Thus if C lies above the axis of abscissas and $x_0 < x$, we find that

$$(x - x_0) p(x_0, x) \leqq h(x_0, x) \leqq (x - x_0) q(x_0, x).$$

If C lies above the axis of abscissas and $x < x_0$, the repeated inequalities just written fail: the signs "$\leqq$" must be replaced by "$\geqq$". To obtain a single formula, we use the familiar function $\operatorname{sign} t$ (equal to $+1$ for $t > 0$ and to -1 for $t < 0$). We can write a single repeated inequality valid when the curve $a_0 a$ lies entirely above the axis of abscissas:

$$(x - x_0) p(x_0, x) \operatorname{sign}(x - x_0) \leqq h(x_0, x) \operatorname{sign}(x - x_0)$$
$$\leqq (x - x_0) q(x_0, x) \operatorname{sign}(x - x_0). \qquad (1)$$

The repeated inequalities (1) are valid also when the curve $a_0 a$ lies entirely below the axis of abscissas: we leave the simple details of the proof to the reader.

Now suppose that the curve $a_0 a$ intersects the axis of abscissas in various points, perhaps an infinite number. Suppose further that the function $f(\xi)$ assumes both positive and negative values in the closed interval $[x_0 x]$. In this case, we do not have a single "quadrilateral" $x_0 x a a_0$, since the curve $a_0 a$ lies partly above and partly below the axis of abscissas. We then add all of the areas that lie below the upper part of the curve $a_0 a$ and the axis of abscissas and subtract from this number the sum of all of the areas that lie above the lower part of the curve $a_0 a$ and below the axis of abscissas. Multiply the resulting number by $\operatorname{sign}(x - x_0)$. We denote this number by $h(x_0, x)$. Once again it is easy to verify the repeated inequalities (1) for this general definition of $h(x_0, x)$.

In the sequel we will regard x_0 as fixed and x as variable. We drop x_0 from our notation and write

$$h(x) = h(x_0, x).$$

It is clear that

$$h(x_0) = h(x_0, x_0) = 0. \tag{2}$$

We will prove next that

$$h'(x) = f(x). \tag{3}$$

Then from (2) and (3) we will infer that

$$h(x) = \int_{x_0}^{x} f(\xi) d\xi. \tag{4}$$

(See relations (2) and (4) in §25.) Thus we will have a geometric interpretation of the definite integral as an area as soon as we have established (3).

In proving (3), it is reasonable to consider the case where the curve $a_0 a$ lies above the axis of abscissas; we keep in mind that all of the formulas we set down are valid in the general case. (Of course strict rigor would demand that we verify this carefully in each case.) To compute the derivative of the function $h(x)$, we consider a quantity ξ that is close to but different from x. Draw the vertical line through the point ξ on the axis of abscissas. Let $\alpha = f(\xi)$ be the point where this line intersects the graph C. We consider the difference

$$h(\xi) - h(x) = h(x_0, \xi) - h(x_0, x).$$

It is geometrically obvious that

$$h(x_0, \xi) - h(x_0, x) = h(x, \xi). \tag{5}$$

This follows from examining the two "quadrilaterals" $x_0 \xi \alpha a_0$ and $x_0 x a a_0$ (once again refer to Fig. 65). Note that the equality (5) holds both for $x < \xi$ and for $x > \xi$. In the inequalities (1), let us write ξ for x and x for x_0. We obtain the inequalities

$$(\xi - x) p(x, \xi) \operatorname{sign}(\xi - x) \leq h(x, \xi) \operatorname{sign}(\xi - x)$$
$$\leq (\xi - x) q(x, \xi) \operatorname{sign}(\xi - x).$$

Divide these inequalities by the positive quantity $(\xi - x) \operatorname{sign}(\xi - x)$ and obtain

$$p(x, \xi) \leq \frac{h(\xi) - h(x)}{\xi - x} \leq q(x, \xi). \tag{6}$$

Since the function $f(x)$ is continuous, the relations

$$\lim_{\xi \to x} p(x, \xi) = \lim_{\xi \to x} q(x, \xi) = f(x)$$

obtain. Thus, taking the limit as $\xi \to x$ in the inequalities (6), we find that

$$h'(x) = f(x).$$

Thus we have proved (3) and so have achieved a geometric interpretation of the definite integral. We have also found a method of computing areas of the form $x_0 x a a_0$ in the cases where we can obtain by hook or by crook an indefinite integral of the function $f(x)$. We give an example.

Example 1. Consider a plane P with coordinate axes as in Fig. 65. In this plane, we consider the parabola with equation

$$2\gamma \xi = y^2,$$

where γ is a positive constant. The axis of this parabola is the positive semi-axis of abscissas. Choose a point x on the positive semi-axis of abscissas. Draw the vertical line through this point x, and let a and a' be the points of intersection of this line with the upper and lower branches of the parabola, respectively. We will compute the area of the region $S(x)$ that is contained within the arc aoa' of the parabola (recall that o is the origin) and the vertical line segment $[aa']$. This region is divided by the axis of abscissas into two parts of equal areas. Let us compute the area of this upper part, which is bounded by the closed interval $[o\,x]$, the vertical line segment $[x\,a]$, and the arc oa of the parabola. The upper branch of the parabola has equation

$$y = +\sqrt{2\gamma\xi}.$$

As shown in (4), therefore, the area of the upper half of our region is

$$h(x) = \int_0^x \sqrt{2\gamma\xi}\, d\xi.$$

We compute with no difficulty:

$$h(x) = \sqrt{2\gamma}\,\tfrac{2}{3} x^{3/2},$$

since for this function $h(x)$ we have

$$h(0) = 0 \quad \text{and} \quad h'(x) = \sqrt{2\gamma x}.$$

Therefore the area $S(x)$ that we wished to compute is

$$S(x) = 2h(x) = \tfrac{4}{3}\sqrt{2\gamma x^3}.$$

§28. The Definite Integral as the Limit
of a Sequence of Finite Sums

In the preceding section, §27, we came very close to a constructive definition of the definite integral. We turn our attention now to this construction. At the same time, we give a formal definition of areas, which we used purely intuitively in §27. We require at the outset the notion of a uniformly continuous function, which we take up first.

Uniform Continuity. Consider a real- or complex-valued function $f(z)$ of a real or complex variable z. We will explain the circumstances under which $f(z)$ is to be considered to be uniformly continuous. Let (z, ζ) be any pair of values for which f is defined. We associate with this pair the two nonnegative numbers

$$|\zeta - z| \quad \text{and} \quad |f(\zeta) - f(z)|.$$

It may happen that for every change of the pair of numbers (z, ζ) under which $|\zeta - z| \to 0$, it also occurs that $|f(\zeta) - f(z)| \to 0$. If this is the case, the function $f(z)$ is called uniformly continuous. More specifically, suppose that for every positive number ε, there is a positive number δ such that for all pairs (z, ζ) where the function f is defined, the inequality

$$|\zeta - z| < \delta \tag{1}$$

implies that

$$|f(\zeta) - f(z)| < \varepsilon. \tag{2}$$

Then the function $f(z)$ is said to be *uniformly continuous*. We write this condition briefly as

$$\lim_{|\zeta - z| \to 0} |f(\zeta) - f(z)| = 0.$$

Suppose that we fix the quantity z in this definition, and let ζ alone vary. We then obtain the definition of continuity of the function at the point z (see §21, (5)).

As we know, a function is said to be continuous if it is continuous at all points where it is defined. It is therefore plain that a uniformly continuous function is continuous. The converse, however, is not in general true (see Example 1).

Suppose now that the set M where the function $f(z)$ is defined is a closed and bounded set. If $f(z)$ is continuous, it is automatically uniformly continuous. We prove this by contradiction. Assume that the function $f(z)$ is continuous but not uniformly continuous. Then there is a positive number ε such that for every positive number δ, one can choose a pair (z, ζ) of points in M for which

$$|z - \zeta| < \delta$$

and
$$|f(\zeta)-f(z)|\geq\varepsilon>0.$$

Let us choose $\delta=\dfrac{1}{n}$. We then find a sequence of pairs of points in M, say

$$z_1,\zeta_1;z_2,\zeta_2;\ldots;z_n,\zeta_n;\ldots$$

such that
$$\lim_{n\to\infty}|\zeta_n-z_n|=0 \tag{3}$$

and
$$|f(\zeta_n)-f(z_n)|\geq\varepsilon>0. \tag{4}$$

The set M is closed and bounded by hypothesis. The points

$$z_1,z_2,\ldots,z_n,\ldots \tag{5}$$

all belong to M. Therefore there is a subsequence of the sequence (5) that converges to a point z_0 in the set M (see §15). For notational convenience we suppose that the sequence itself converges to z_0:

$$\lim_{n\to\infty}z_n=z_0. \tag{6}$$

From (3) and (6) we see that

$$\lim_{n\to\infty}\zeta_n=z_0. \tag{7}$$

To see this, note that

$$|\zeta_n-z_0|\leq|\zeta_n-z_n|+|z_n-z_0|.$$

Take the limit of the right side of this inequality as $n\to\infty$: this gives (7). The function $f(z)$ is continuous everywhere, also by hypothesis, and so it is continuous at the point z_0. This implies that

$$\lim_{n\to\infty}f(z_n)=f(z_0) \tag{8}$$

and that

$$\lim_{n\to\infty}f(\zeta_n)=f(z_0). \tag{9}$$

We also have

$$|f(\zeta_n)-f(z_n)|\leq|f(\zeta_n)-f(z_0)|+|f(z_n)-f(z_0)|.$$

Take the limit as $n\to\infty$ in the right side of this inequality and apply (8) and (9). We find that

$$\lim_{n\to\infty}|f(\zeta_n)-f(z_n)|=0,$$

which contradicts (4).

We have thus proved that a continuous function on a closed bounded set is uniformly continuous.

We proceed to our constructive definition of the definite integral

$$h(x) = \int_{x_0}^{x} f(\xi)\, d\xi \tag{10}$$

of a real-valued continuous function $f(\xi)$ of a real variable ξ, defined on a connected set of real numbers. We will define this integral by approximating to it.

Just as in §27, we denote the minimum and maximum values of the function $f(\xi)$ on a closed interval $[x_0 x]$ by the symbols

$$p(x_0, x) \quad \text{and} \quad q(x_0, x) \quad \text{respectively.} \tag{11}$$

We also write

$$\varepsilon(x_0, x) = q(x_0, x) - p(x_0, x). \tag{12}$$

The nonnegative number $\varepsilon(x_0, x)$ measures the amount by which $f(\xi)$ varies in the closed interval $[x_0 x]$. For any two points μ and v of the closed interval $[x_0 x]$, it is clear that

$$|f(\mu) - f(v)| \leq \varepsilon(x_0, x). \tag{13}$$

Subdivisions of a Closed Interval. We divide the closed interval $[x_0 x]$ into subintervals by a monotone finite sequence of points

$$X = (x_0, x_1, \ldots, x_i, \ldots, x_k) \quad \text{where } x_k = x. \tag{14}$$

If $x_0 < x$, monotonicity means that

$$x_0 < x_1 < \cdots < x_i < \cdots < x_k; \tag{15}$$

if $x_0 > x$, it means that

$$x_0 > x_1 > \cdots > x_i > \cdots > x_k. \tag{16}$$

In both cases, we have

$$\text{sign}\,(x_i - x_{i-1}) = \text{sign}\,(x - x_0). \tag{17}$$

The sequence of points (14) divides the closed interval $[x_0 x]$ into k subintervals $[x_{i-1} x_i]$. We call X as in (14) *a subdivision of the interval* $[x_0 x]$. We note also the trivial subdivision

$$X = (x_0, x), \tag{18}$$

that introduces no division points at all. Our constructive definition of the integral (10) depends upon a consideration of an infinite number of subdivisions X. Thus we will think of the subdivision X as a variable quantity, the values of which are arbitrary subdivisions of the closed interval $[x_0 x]$. Neither the number k of subintervals nor the endpoints of the subdivisions is constant. On the variable quantity X, we will define a number of needed numerical functions. For example, let $\delta(X)$ be the largest of the positive numbers

$$|x_i - x_{i-1}| \text{ as } i \text{ runs from 1 to } k.$$

The number $\delta(X)$ is a computable function of the subdivision X. It measures the "fineness" of the subdivision X: the smaller $\delta(X)$ is, the finer is the subdivision X. Suppose that X is the subdivision of $[x_0 x]$ into k subdivisions of equal lengths. Then we have

$$\delta(X) = \frac{|x - x_0|}{k}.$$

From this we see that $[x_0 x]$ admits subdivisions X for which $\delta(X)$ is arbitrarily small, since $\frac{1}{k}|x - x_0|$ goes to 0 as $k \to \infty$.

Throughout the subsequent discussion of the definition of (10), we think of $f(\xi)$ as a fixed continuous real-valued function. Let $\varepsilon(X)$ be the greatest of the numbers $\varepsilon(x_{i-1}, x_i)$, where $\varepsilon(x_{i-1}, x_i)$ is defined by (12). The nonnegative number $\varepsilon(X)$ (which is zero for all X only for a constant function $f(\xi)$) is also a function of the subdivision X, which can be computed once X is known. Suppose that μ and ν are any two points that lie in a single subinterval $[x_{i-1} x_i]$. It is clear that

$$|f(\mu) - f(\nu)| \le \varepsilon(X).$$

The function $f(\xi)$ being continuous, it is uniformly continuous on the closed and bounded interval $[x_0 x]$. For every positive number ε, therefore, there exists a positive number δ such that for all subdivisions X with $\delta(X) < \delta$, the inequality $\varepsilon(X) < \varepsilon$ holds. In other words, suppose that X varies in such a way that as X varies, the number $\delta(X)$ goes to 0. Then the number $\varepsilon(X)$ also goes to 0. We write this as

$$\lim_{\delta(X) \to 0} \varepsilon(X) = 0. \tag{19}$$

We introduce yet another function of the subdivision X. We write

$$\sigma(X) = (x_1 - x_0) f(\xi_1) + \cdots + (x_i - x_{i-1}) f(\xi_i)$$
$$+ \cdots + (x_k - x_{k-1}) f(\xi_k), \tag{20}$$

where ξ_i is an arbitrary point in the subinterval $[x_{i-1} x_1]$ for $i = 1, 2, \ldots, k$. The number $\sigma(X)$ is thus not a function of X alone, but also of the selection of the points

$$\xi_1, \xi_2, \ldots, \xi_i, \ldots, \xi_k. \tag{21}$$

To be accurate, we should write

$$\sigma(X; \xi_1, \xi_2, \ldots, \xi_i, \ldots, \xi_k).$$

We can also think of $\sigma(X)$ as being a multiple-valued function of X, which can assume all of the possible values that come up when we choose all possible sequences (21). (Recall that ξ_i must belong to the subinterval $[x_{i-1}, x_i]$.) All properties of the function $\sigma(X)$ that will be proved and used below are true of all of its values. That is, they are independent of the choice of the sequence (21). To shorten our notation, we will write the expression (20) as

$$\sigma(X) = \sum_{i=1}^{k} (x_i - x_{i-1}) f(\xi_i). \tag{22}$$

We will constantly in what follows use the "sigma notation" $\sum$, which indicates that the sum on the right side is to be taken over all values $i = 1, 2, \ldots, k$. The formula (22) makes sense both for $x > x_0$ and for $x < x_0$. For $x = x_0$, however, the interval $[x_0\, x]$ reduces to the single point x_0 and admits no subdivisions at all. In this case, we formally write

$$\sigma(X) = 0. \tag{23}$$

The finite sums $\sigma(X)$ are approximants for the integral (10). The approximation grows better and better as $\delta(X)$ gets closer and closer to 0.

Definition of the Integral. We will show that if the quantity X varies in such a way that its fineness $\delta(X)$ goes to 0, then the quantity $\sigma(X)$ converges to a specific limit, which we will call $h(x)$. We write this assertion as a formula:

$$\lim_{\delta(X) \to 0} \sigma(X) = h(x). \tag{24}$$

In precise terms, (24) means the following. For every positive number ε, there exists a positive number δ such that if

$$\delta(x) < \delta,$$

then

$$|\sigma(X) - h(x)| < \varepsilon. \tag{25}$$

We proceed to our proof. First, for the trivial subdivision (18), we have

$$\sigma(x_0, x) = (x - x_0) f(\xi),$$

where ξ is an arbitrary point in the closed interval $[x_0 \, x]$. Let us prove that for an arbitrary subdivision X, the inequality

$$|\sigma(x_0, x) - \sigma(X)| \leq |x - x_0| \sigma(x_0, x) \qquad (26)$$

obtains. We write

$$\sigma(x_0, x) = \sum_{i=1}^{k} (x_i - x_{i-1}) f(\xi). \qquad (27)$$

Subtract (27) from (22) and obtain

$$\sigma(x_0, x) - \sigma(X) = \sum_{i=1}^{k} (x_i - x_{i-1})(f(\xi) - f(\xi_i)). \qquad (28)$$

From (13), we have

$$|f(\xi) - f(\xi_i)| \leq \varepsilon(x_0, x). \qquad (29)$$

Combining (28) and (29), we find that

$$|\sigma(x_0, x) - \sigma(X)| \leq \sum_{i=1}^{k} |x_i - x_{i-1}| \varepsilon(x_0, x) = |x - x_0| \varepsilon(x_0, x).$$

This is exactly the inequality (26).

Now suppose that we have a monotone sequence, as defined in (15) and (16), which we write

$$X' = (x'_0, x'_1, \ldots, x'_i, \ldots, x'_l), \qquad \text{where} \quad x'_0 = x_0 \quad \text{and} \quad x'_l = x.$$

Suppose that every point of subdivision in the definition (14) of X is one of the points of subdivision of X'. We then say that X' is *a refinement of* X. We will show that if X' is a refinement of X, then

$$|\sigma(X) - \sigma(X')| \leq |x - x_0| \varepsilon(X). \qquad (30)$$

For a given subinterval $[x_{i-1} \, x_i]$, the points of the subdivision X' that lie in this subinterval form a subdivision of $[x_{i-1} \, x_i]$ into subsubintervals. We denote this subdivision of $[x_{i-1} \, x_i]$ by X'_i. We apply (26) to $[x_{i-1} \, x_i]$ and X'_i. This yields

$$|\sigma(x_{i-1}, x_i) - \sigma(X'_i)| \leq |x_i - x_{i-1}| \varepsilon(x_{i-1}, x_i) \leq |x_i - x_{i-1}| \varepsilon(X). \qquad (31)$$

It is clear that

$$\sigma(X') = \sum_{i=1}^{k} \sigma(X'_i) \qquad (32)$$

and

$$\sigma(X) = \sum_{i=1}^{k} \sigma(x_{i-1}, x_i). \qquad (33)$$

Subtract (32) from (33) and take account of (31). This gives

$$|\sigma(X) - \sigma(X')| = \left| \sum_{i=1}^{k} [\sigma(x_{i-1}, x_i) - \sigma(X_i')] \right|$$

$$\leq \sum_{i=1}^{k} |x_i - x_{i-1}| \, \varepsilon(X) = |x - x_0| \, \varepsilon(X).$$

This is the inequality (30).

Suppose now that X and X'' are any two subdivisions of the closed interval $[x_0 \, x]$. We will prove that

$$|\sigma(X) - \sigma(X'')| \leq |x - x_0| (\varepsilon(X) + \varepsilon(X'')). \tag{34}$$

We consider all of the numbers that appear in at least one of the sequences X and X''. Arrange this set of numbers in monotone order (increasing if $x_0 < x$ and decreasing if $x_0 > x$). Let X' be the resulting subdivision of $[x_0 \, x]$. Plainly X' is a common refinement of the subdivisions X and X''. Along with (30), we thus have the inequality

$$|\sigma(X'') - \sigma(X')| \leq |x - x_0| \, \varepsilon(X''). \tag{35}$$

From (30) and (35), we obtain (34) by adding.

We can now construct the number $h(x)$ having the property (24). We choose any sequence at all of subdivisions of $[x_0 \, x]$,

$$X_1, X_2, \ldots, X_n, \ldots$$

having the property that

$$\lim_{n \to \infty} \delta(X_n) = 0. \tag{36}$$

We will prove that the sequence of numbers

$$\sigma(X_1), \sigma(X_2), \ldots, \sigma(X_n), \ldots \tag{37}$$

converges. We denote its limit by $h(x)$:

$$\lim_{n \to \infty} \sigma(X_n) = h(x). \tag{38}$$

We will prove that the sequence (37) is a Cauchy sequence of numbers (see §14). We make an estimate of the difference $|\sigma(X_r) - \sigma(X_s)|$, based on (34). The estimate (34) gives us

$$|\sigma(X_r) - \sigma(X_s)| \leq |x - x_0| (\varepsilon(X_r) + \varepsilon(X_s)). \tag{39}$$

By our hypothesis (36), the numbers $\delta(X_n)$ go to zero. It follows from (19) that the numbers $\varepsilon(X_n)$ also go to zero, and so the right side of (39) is arbitrarily small for all sufficiently large positive integers r and s. That is, (37) is a Cauchy sequence of numbers and so has a limit, which, as noted in (38), we write as $h(x)$.

We can now show that the quantity $h(x)$ satisfies (24). We have

$$|\sigma(X)-h(x)|\leq|\sigma(X)-\sigma(X_n)|+|\sigma(X_n)-h(x)|$$
$$\leq|x-x_0|(\varepsilon(X)+\varepsilon(X_n))+|\sigma(X_n)-h(x)|.$$

In this sequence of inequalities, take the limit as $n\rightarrow\infty$. We find that

$$|\sigma(X)-h(x)|\leq|x-x_0|\,\varepsilon(X). \tag{40}$$

This inequality and (19) give us the equality (24).

We can now give our final definition of the definite integral:

$$\int_{x_0}^{x}f(\xi)\,d\xi=\lim_{\delta(X)\rightarrow0}\sigma(X).$$

We set down an important estimate for the size of the definite integral:

$$(x-x_0)p(x_0,x)\operatorname{sign}(x-x_0)\leq\operatorname{sign}(x-x_0)\int_{x_0}^{x}f(\xi)\,d\xi$$
$$\leq(x-x_0)q(x_0,x)\operatorname{sign}(x-x_0). \tag{41}$$

(See also the inequalities (44) in §25.) To prove (41), we consider an arbitrary subdivision X of the closed interval $[x_0\,x]$. We have

$$p(x_0,x)\leq f(\xi_i)\leq q(x_0,x) \qquad \text{for } i=1,2,...,k.$$

Multiplying this pair of inequalities by the positive number

$$(x_i-x_{i-1})\operatorname{sign}(x_i-x_{i-1})$$

and adding for all $i=1,2,...,k$, we obtain the inequalities

$$(x-x_0)p(x_0,x)\operatorname{sign}(x-x_0)\leq\operatorname{sign}(x-x_0)\sigma(X)$$
$$\leq(x-x_0)q(x_0,x)\operatorname{sign}(x-x_0). \tag{42}$$

In the inequalities (42), we may take the limit as $\delta(X)\rightarrow0$ and obtain the inequalities (41).

It is now time to show that the function $h(x)$ satisfies the conditions

$$h(x_0) = 0 \tag{43}$$

and

$$h'(x) = f(x). \tag{44}$$

That is, we must show that $h(x)$ is the definite integral of $f(x)$ as the term was defined in §25 (see §25, conditions (2) and (4)). Condition (43) follows at once from (23). To prove (44), we first divide the interval $[x_0 \, x]$ into two subintervals by a point x^1 such that x^1 lies strictly between x_0 and x. We rename the point x_0 as x^0 and the point x as x^2. Consider any subdivision X^1 of the interval $[x^0 \, x^1]$ and any subdivision X^2 of the interval $[x^1 \, x^2]$. If we write the points of the subdivision X^2 in their order *after* the points of the subdivision X^1 in their order, we obtain a subdivision of the interval $[x^0 \, x^2]$, which we denote by the symbol $X^1 \, X^2$. It is clear that

$$\sigma(X^1) + \sigma(X^2) = \sigma(X^1 \, X^2). \tag{45}$$

Going to the limit in (45) as $\delta(X^1) \to 0$ and $\delta(X^2) \to 0$, we find that

$$\int_{x^0}^{x^1} f(t)\,dt + \int_{x^1}^{x^2} f(t)\,dt = \int_{x^0}^{x^2} f(t)\,dt. \tag{46}$$

In computing the derivative $h'(x)$, we follow our usual rule of choosing a point ξ that is close to but distinct from x and forming the difference $h(\xi) - h(x)$. Suppose first that $x \neq x_0$. We may suppose that $|\xi - x| < |x - x_0|$, since we are concerned only with values of ξ that are close to x. If $x_0 < x < \xi$ or $\xi < x < x_0$, (46) gives us

$$\int_{x_0}^{x} f(t)\,dt + \int_{x}^{\xi} f(t)\,dt = \int_{x_0}^{\xi} f(t)\,dt,$$

so that

$$h(\xi) - h(x) = \int_{x}^{\xi} f(t)\,dt. \tag{47}$$

If $x_0 < \xi < x$ or $x < \xi < x_0$, (46) gives

$$\int_{x_0}^{\xi} f(t)\,dt + \int_{\xi}^{x} f(t)\,dt = \int_{x_0}^{x} f(t)\,dt,$$

so that

$$h(\xi) - h(x) = - \int_{\xi}^{x} f(t)\,dt. \tag{48}$$

It is easy to see from the definition (38) that

$$\int_{\xi}^{x} f(t)\,dt = - \int_{x}^{\xi} f(t)\,dt. \tag{49}$$

Combining (48) and (49), we see that (47) holds for all values of ξ that are sufficiently close to x. Since $h(x_0)=0$, (47) holds also for $x=x_0$.

We now go back to the inequalities (41). In these inequalities, replace x by ξ, x_0 by x, and the integral by $h(\xi)-h(x)$, as in (47). We find that

$$(\xi-x)\,p(x,\xi)\,\text{sign}(\xi-x)\leqq\text{sign}(\xi-x)[h(\xi)-h(x)]$$
$$\leqq(\xi-x)\,q(x,\xi)\,\text{sign}(\xi-x). \tag{50}$$

Divide all three terms in (50) by the positive number $(\xi-x)\,\text{sign}(\xi-x)$, obtaining

$$p(x,\xi)\leqq\frac{h(\xi)-h(x)}{\xi-x}\leqq q(x,\xi). \tag{51}$$

Since $f(x)$ is a continuous function, the numbers $p(x,\xi)$ and $q(x,\xi)$ are arbitrarily close to $f(x)$ for all ξ that are sufficiently close to x. Thus we may take the limit as $\xi\to x$ in (51) to find that

$$f(x)\leqq h'(x)\leqq f(x),$$

relations that are obviously equivalent to (44). This completes our proof.

Piecewise Continuous Functions. We now give a generalization of the notion of definite integral, defined so far only for continuous real-valued functions of a real variable. Suppose that $f(x)$ is a real-valued function of a real variable that is discontinuous at some point a where $f(x)$ is defined. We may call $x=a$ a *point of discontinuity* of $f(x)$. We will describe the simplest sort of discontinuity of $f(x)$, called a *discontinuity of the first kind*. Suppose that the following conditions hold.

1) In some neighborhood of a, a is the only point of discontinuity of $f(x)$. That is, there is a positive number ε such that $f(x)$ is continuous at all points x such that $a-\varepsilon<x<a$ or $a<x<a+\varepsilon$.

2) As x approaches a through values less than a, the values of $f(x)$ approach a limit, which we denote by $f(a-0)$. We write:

$$\lim_{\substack{x\to a\\x<a}}f(x)=f(a-0).$$

As x approaches a through values greater than a, $f(x)$ also approaches a limit, which we denote by $f(a+0)$. We write

$$\lim_{\substack{x\to a\\x>a}}f(x)=f(a+0).$$

Naturally we must have $f(a-0)\neq f(a)$ or $f(a+0)\neq f(a)$ if $f(x)$ is discontinuous at a. If $f(a-0)=f(a+0)$, then $f(x)$ has a *removable discontinuity at a.* If $f(a-0)\neq f(a+0)$, and condition 1) on p. 247 holds, we say that $f(x)$ has *a discontinuity of the first kind at a.*

As a rule, the value assigned to $f(a)$ is of little importance if a is a point of discontinuity of the first kind. Suppose that $f(x)$ is a function having discontinuities only of the first kind and also such that on every interval of finite length, there are only a finite number of discontinuities of $f(x)$. Then the function $f(x)$ is said to be *piecewise continuous.*

It is simple to define the definite integral

$$\int_{x_0}^{x} f(\xi)\,d\xi$$

of a piecewise continuous function. For definiteness let us suppose that $x_0 < x$. Suppose that $f(x)$ is continuous at all points of the interval $[x_0\,x]$ except possibly at its endpoints, where f may be continuous or have a discontinuity of the first kind. If x_0 is a point of discontinuity of the first kind, we may redefine $f(x_0)$ to be the number $f(x_0+0)$. This makes the function continuous at x_0. If x is a point of discontinuity of the first kind, we define $f(x)$ to be $f(x-0)$. The function f is then continuous throughout the entire closed interval $[x_0\,x]$ and so the integral of f is defined. Now suppose that $[x_0\,x]$ is an arbitrary interval. We can write $[x_0\,x]$ as the union of a finite number of intervals, as in (14) and (15), within each of which f is continuous (with redefinitions as needed at the points of discontinuity). We define

$$\int_{x_0}^{x} f(\xi)\,d\xi$$

as the sum of the integrals of f taken over each of these subintervals. For $x_0 > x$, the definition is similar, or we may use the identity (49).

Piecewise Smooth Functions. Let $h(x)$ be a continuous function such that its derivative $h'(x)$ is defined except at a number of points such that every interval of finite length contains only a finite number of points where $h'(x)$ is not defined. Suppose also that $h'(x)$ is a piecewise continuous function. Then we say that $h(x)$ is *piecewise smooth.* Now let $f(x)$ be a piecewise continuous function. It is easy to see that its integral,

$$h(x)=\int_{x_0}^{x} f(\xi)\,d\xi,$$

is a piecewise smooth function and that $h'(x)=f(x)$ for all x such that $f(x)$ is continuous at x.

Integration of a Complex-valued Function $f(x)$ of a Real Variable x. Let us write $f(x)$ in the form

$$f(x)=\varphi(x)+i\,\psi(x),$$

where $\varphi(x)$ and $\psi(x)$ are real-valued functions of the real variable x. Let X be a decomposition of the interval $[x_0\,x]$ over which we wish to integrate. We write a sum of type σ for $f(x)$:

$$\sigma(X) = \sum_{j=1}^{k} (x_j - x_{j-1}) f(\xi_j), \tag{52}$$

where ξ_j is a point in the interval $[x_{j-1}\,x_j]$ for $j = 1, 2, ..., k$. We also write

$$\sigma_1(X) = \sum_{j=1}^{k} (x_j - x_{j-1}) \varphi(\xi_j)$$

and

$$\sigma_2(X) = \sum_{j=1}^{k} (x_j - x_{j-1}) \psi(\xi_j).$$

It is obvious that

$$\sigma(X) = \sigma_1(X) + i\,\sigma_2(X). \tag{53}$$

Suppose that $f(x)$ is continuous. Then the real-valued functions $\varphi(x)$ and $\psi(x)$ are also continuous, and we have

$$\lim_{\delta(X) \to 0} \sigma_1(X) = \int_{x_0}^{x} \varphi(\xi)\,d\xi$$

and

$$\lim_{\delta(X) \to 0} \sigma_2(X) = \int_{x_0}^{x} \varphi(\xi)\,d\xi.$$

The formulas (52) and (53) show that the limit of $\sigma(X)$ as $\delta(X) \to 0$ exists and can be taken as the definition of the integral. We thus find that

$$\int_{x_0}^{x} f(\xi)\,d\xi = \lim_{\delta(X) \to 0} \sigma(X) = \int_{x_0}^{x} \varphi(\xi)\,d\xi + i \int_{x_0}^{x} \psi(\xi)\,d\xi. \tag{54}$$

If $f(x)$ is not continuous but piecewise continuous, so are the functions $\varphi(x)$ and $\psi(x)$ (although not perhaps at the same points). We can then define the integral of $f(x)$ just as we did for piecewise continuous real-valued functions. Once again the identity (54) holds.

For the functions $\varphi(x)$ and $\psi(x)$, we have

$$\int_{x_0}^{x_0} \varphi(\xi)\,d\xi = 0, \quad \int_{x_0}^{x_0} \psi(\xi)\,d\xi = 0,$$

as well as

$$\frac{d}{dx} \int_{x_0}^{x} \varphi(\xi)\,d\xi = \varphi(x), \quad \frac{d}{dx} \int_{x_0}^{x} \psi(\xi)\,d\xi = \psi(x).$$

The same identities therefore hold for the complex-valued function $f(x)$:

$$\int_{x_0}^{x_0} f(\xi)\,d\xi = 0 \quad \text{and} \quad \frac{d}{dx}\int_{x_0}^{x} f(\xi)\,d\xi = f(x). \tag{55}$$

We have accordingly proved that the integral

$$\int_{x_0}^{x} f(\xi)\,d\xi \tag{56}$$

of our complex-valued function $f(x)$ of a real variable x, defined as in (54), satisfies the defining conditions of the definite integral as in §25. Therefore all that we established in §25 for the definite integral is true for the integral (56), in particular the identity (10) of §25

$$\int_{x_0}^{x_1} f(\xi)\,d\xi + \int_{x_1}^{x_2} f(\xi)\,d\xi = \int_{x_0}^{x_2} f(\xi)\,d\xi. \tag{57}$$

(See also (46) for a special case.)

Suppose that the integrand $f(\xi)$ in (56) satisfies the condition

$$|f(\xi)| \leq m$$

for all ξ in the closed interval $[x_0\, x]$. We then have

$$\left|\int_{x_0}^{x} f(\xi)\,d\xi\right| \leq m\,|x - x_0|. \tag{58}$$

This follows at once from the fact that the integral (56) is the limit of the sums $\sigma(X)$ defined in (52).

Example. We give an example of a continuous function that is not uniformly continuous: $f(x) = \dfrac{1}{x}$ for all x such that $0 < x \leq 1$. This function is obviously continuous, but it fails to be uniformly continuous, since

$$f(\xi) - f(x) = \frac{1}{\xi} - \frac{1}{x} = \frac{x - \xi}{\xi x}$$

can be made arbitrarily large for $|x - \xi| < \delta$ if x and ξ are sufficiently close to zero.

The function $f(x) = e^x$, defined for all real numbers x, is a slightly different example of a continuous but not uniformly continuous function. For any number a, e^x is uniformly continuous on the set $-\infty < x \leq a$. However, the identity

$$e^{x+\delta} - e^x = e^x(e^\delta - 1)$$

shows that for every positive number δ, no matter how small, we can choose x so large that $e^{x+\delta} - e^x$ is arbitrarily large.

§29. Area and Curve Length

In §27, we considered the area of a plane figure bounded by three straight line segments and a piece of the graph of a continuous function as being an entity that needed no further explanation. Every right-thinking person, it would seem, must believe that areas can be measured. Nevertheless, "facts" that are evident in this sense can sometimes be plain wrong. We must therefore explain with care just what we mean by the area of a region in the plane that is bounded by some closed curve. The problem of defining the length of a curve is much the same. To avoid possible pitfalls, we must, as mathematicians say, give a rigorous definition of these concepts. We will interpret both of these notions as integrals.

Area. Consider a continuous real-valued function $f(\xi)$ of a real variable ξ. Let C be the graph of this function in our coordinate plane. We re-use the notation employed at the beginning of §27, as well as Fig. 65. For the sake of definiteness, we will suppose that the part of the curve C lying between x_0 and x is entirely above the axis of abscissas. We set ourselves the problem of defining the area of the plane region bounded by the line segments $[x_0 a_0]$, $[x_0 x]$, $[x a]$, and the curve $a_0 a$. See Fig. 65 for a sketch. In the hope that we will succeed, we denote this area by the symbol S.

As in previous work, we use special notation for the minimum and maximum values of our continuous function $f(\xi)$. Let μ and v be two points on the axis of abscissas. Let $p(\mu, v)$ be the minimum of $f(\xi)$ on the closed interval $[\mu v]$ and $q(\mu, v)$ the maximum of $f(\xi)$ on this interval. We proved in §23 that these values are assumed:

$$p(\mu, v = f(\xi') \quad \text{and} \quad q(\mu, v) = f(\xi'')$$

for certain points ξ' and ξ'' of the closed interval $[\mu v]$. (The points ξ' and ξ'' need not be unique, of course.)

We now choose an arbitrary subdivision of of the closed interval $[x_0 x]$ as in §28, (14) (we take $x_0 < x$):

$$X = (x_0, x_1, \ldots, x_i, \ldots, x_k) \quad \text{with } x_k = x.$$

Let ξ'_i be a point in the subinterval $[x_{i-1} x_i]$ where the function $f(\xi)$ assumes its least value in this subinterval. Let ξ''_i be the same with respect to the greatest value. We write two sums:

$$\sigma'(X) = \sum_{i=1}^{k} (x_i - x_{i-1}) f(\xi_i')$$

and

$$\sigma''(X) = \sum_{i=1}^{k} (x_i - x_{i-1}) f(\xi_i'').$$

The sums $\sigma'(X)$ and $\sigma''(X)$ are possible values for the multiple-valued function $\sigma(X)$ (see §28). We therefore have

$$\lim_{\delta(X) \to 0} \sigma'(X) = \int_{x_0}^{x} f(\xi) d\xi \tag{1}$$

and

$$\lim_{\delta(X) \to 0} \sigma''(X) = \int_{x_0}^{x} f(\xi) d\xi. \tag{2}$$

In Fig. 66, we give a sketch of the behavior of $\sigma'(X)$ and $\sigma''(X)$. Let P_i be the rectangle with base $[x_{i-1} x_i]$ and altitude $f(\xi_i')$. Let Q_i be the rectangle with the same base and altitude $f(\xi_i'')$. The area of the rectangle P_i, as we learned in elementary geometry, is $(x_i - x_{i-1}) f(\xi_i')$, while the area of Q_i is $(x_i - x_{i-1}) f(\xi_i'')$. All of the rectangles P_i are contained within the region $x_0 \, x \, a \, a_0$, and so the sum of their areas (under any reasonable definition of area) does not exceed the area S of the region $x_0 \, x \, a \, a_0$. Similarly, if we add up all of the rectangles Q_i, we get a plane region that contains the region $x_0 \, x \, a \, a_0$, and so the sum of the areas of the rectangles Q_i must be greater than or equal to the area S. That is, we have

$$\sigma'(X) \leqq S \leqq \sigma''(X). \tag{3}$$

It is thus reasonable to regard both $\sigma'(X)$ and $\sigma''(X)$ as approximations, from below and above, respectively, for the area S. From (1) and (2), we see that both of these quantities have the same limit as $\delta(X) \to 0$. We therefore define the area S as their common limit:

$$S = \int_{x_0}^{x} f(\xi) d\xi.$$

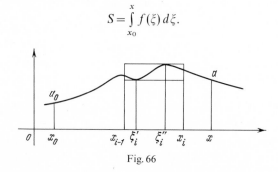

Fig. 66

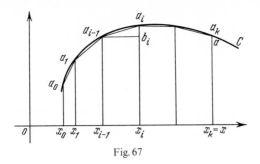

Fig. 67

Thus a formal definition of area leads to the same result as what our intuition suggests.

The Length of a Curve. Consider next a real-valued function $f(x)$ of a real variable having a continuous derivative. Let x_0 and x be two points of the axis of abscissas with $x_0 < x$. Let X be an arbitrary subdivision as in §28, (14), of the closed interval $[x_0 \, x]$. Again we denote the graph of the function $f(x)$ between the limits x_0 and x by C, as in Fig. 67. For $i = 0, 1, \ldots, k$, let a_i be the point $(x_i, f(x_i))$ where the vertical line through x_i on the axis of abscissas meets the curve C. We construct the broken line

$$a_0, a_1, \ldots, a_k = a$$

made up of the line segments $[a_0 \, a_1], [a_1 \, a_2], \ldots, [a_{k-1} \, a_k]$. Let $l(a_{i-1}, a_i)$ be the length of the line segment $[a_{i-1} \, a_i]$. Let $l(X)$ denote the length of the broken line a, that is,

$$l(X) = \sum_{i-1}^{k} l(a_{i-1}, a_i). \tag{4}$$

It is easy to show that if $x_i - x_{i-1} \to 0$, then we also have $l(a_{i-1}, a_i) \to 0$. We define the length $s(x_0, x)$ of the curve C as

$$s(x_0, x) = \lim_{\delta(X) \to 0} l(X).$$

We will show that this limit exists and that it can be written as a definite integral. Draw a horizontal line through the point a_{i-1} on the curve C, and write b_i for the point where this line intersects the vertical line through x_i: a sketch appears in Fig. 67. The theorem of Pythagoras shows that

$$l(a_{i-1}, a_i) = \sqrt{(l(a_{i-1}, b_i))^2 + (l(b_i, a_i))^2}.$$

Since

$$l(a_{i-1}, b_i) = x_i - x_{i-1} \quad \text{and} \quad l(b_i, a_i) = |f(x_i) - f(x_{i-1})|,$$

we may write

$$l(a_{i-1}, a_i) = \sqrt{(x_i - x_{i-1})^2 + (f(x_i) - f(x_{i-1}))^2}.$$

Lagrange's formula (see §23, (20)) shows that

$$f(x_i) - f(x_{i-1}) = f'(\xi_i)(x_i - x_{i-1})$$

for some point ξ_i in the interval $[x_{i-1} x_i]$. Thus the length $l(x_0, x)$ of our broken line is

$$l(X) = \sum_{i=1}^{k} (x_i - x_{i-1}) \sqrt{1 + (f'(\xi_i))^2}.$$

The right side of the last equality is a sum of type σ (see §28) for the function $\sqrt{1 + (f'(\xi))^2}$. Therefore the limit $s(x_0, x)$ exists and is equal to

$$s(x_0, x) = \int_{x_0}^{x} \sqrt{1 + (f'(\xi))^2} \, d\xi. \tag{5}$$

§30. The Length of a Curve Given in Parametric Form

Definite integrals are used to describe and even to define a vast array of quantities met with in mathematics itself and also in applications of mathematics to the natural sciences. The integral in question is frequently encountered as the limit of finite sums of the sort that we studied in §28. These sums may, it is true, be somewhat more general than those we considered in §28. To clarify the connection between these more general sums and those we discussed in §28, we will now consider sums that are so to say "infinitely small", that is, that go to zero, as the subdivision X of our interval grows finer and finer.

We fix a closed interval $[x_0 x]$ on the axis of abscissas. Suppose that for every subinterval $[\mu v]$ of $[x_0 x]$ we have a real or complex number $\rho(\mu, v)$ having the property that $\rho(\mu, v)$ goes to zero as $|\mu - v|$ goes to zero. Specifically, we suppose that for every positive number ε, there is a positive number δ such that the inequality

$$|\mu - v| < \delta \tag{1}$$

implies that

$$|\rho(\mu, v)| < \varepsilon. \tag{2}$$

We write this briefly by the expression

$$\lim_{|v - \mu| \to 0} \rho(\mu, v) = 0. \tag{3}$$

A function $\rho(\mu, v)$ of subintervals $[\mu v]$ that enjoys this property will be called *infinitely small*.

Suppose now that we have a subdivision

$$X = (x_0, x_1, \ldots, x_i, \ldots, x_k) \quad \text{with } x_k = x$$

of the interval $[x_0, x]$ as in §28, (14). For each such subdivision, we define

$$\rho(X) = \sum_{i=1}^{k} (x_i - x_{i-1}) \rho(x_{i-1}, x_i),$$

where $\rho(\mu, v)$ is any infinitely small function of subintervals.

We will show that $\rho(X) \to 0$ as the subdivision X grows finer and finer. We write this briefly as

$$\lim_{\delta(X) \to 0} \rho(X) = 0. \tag{4}$$

To prove this, we take an arbitrary positive number ε and define δ so that the relation (1) implies (2). Let X be any subdivision of $[x_0\, x]$ for which $\delta(X) < \delta$. From (2) we find that

$$|\rho(x_{i-1}, x_i)| < \varepsilon$$

for all adjacent points of the subdivision X. For the function $\rho(X)$ we obviously have

$$|\rho(X)| \le \sum_{i=1}^{k} |x_i - x_{i-1}| \cdot |\rho(x_{i-1}, x_i)| \le \sum_{i=1}^{k} |x_i - x_{i-1}| \varepsilon = |x - x_0| \varepsilon.$$

Since ε is an arbitrary positive number, (4) is proved.

Piecewise Smooth Curves. Let $\varphi(t)$ and $\psi(t)$ be real-valued function of a real variable t. Writing

$$x = \varphi(t) \quad \text{and} \quad y = \psi(t), \tag{5}$$

we obtain a description in terms of the parameter t of a curve in the coordinate plane, traced out by the points

$$(x, y) = (\varphi(t), \psi(t)) \tag{6}$$

as t runs through some interval in the real number system, either increasing or decreasing. If both of the functions $\varphi(t)$ and $\psi(t)$ are piecewise smooth (see §28), we say that the curve (6) is *piecewise smooth*, provided that

$$(\varphi'(t))^2 + (\psi'(t))^2 > 0$$

for all values of the parameter t, barring the finite number of values where one of these derivatives is undefined. A point $(\varphi(t), \psi(t))$ on the curve where at least one derivative $\varphi'(t)$ or $\psi'(t)$ is undefined is called *a corner point of the curve.* Our curve is divided by its corner points into a finite number of subcurves, each of which is a smooth curve. We will suppose also that as we approach a corner point, it is not the case that both of the derivatives $\varphi'(t)$ and $\psi'(t)$ have limit zero. A point that is not a corner point is called *a smooth point.* To give a rigorous definition of the length of a piecewise smooth curve, and also to calculate it, it suffices to define and calculate this length for each of the smooth subcurves that comprise it. We then add the results. We will therefore limit ourselves to defining the length of a curve defined by a parameter interval $[t_0\, t]$ in which the curve is smooth.

Let T be a subdivision of $[t_0\, t]$ as in §28, (14):

$$T = (t_0, t_1, \ldots, t_i, \ldots, t_k) \quad \text{with } t_k = t.$$

We will of course suppose that this subdivision is monotone. Let α_i be the point $(\varphi(t_i)\,\psi(t_i))$ on the curve. The sequence of points

$$\alpha_0, \alpha_1, \ldots, \alpha_i, \ldots, \alpha_k$$

defines a broken line, consisting of the line segments $[\alpha_{i-1}\,\alpha_i]$, $i = 1, 2, \ldots, k$. Let $l(\alpha_{i-1}, \alpha_i)$ be the length of the line segment $[\alpha_{i-1}\,\alpha_i]$. We will consider this length as positive if $t_{i-1} < t_i$ and negative if $t_{i-1} > t_i$. We define the length of the entire broken line by

$$l(T) = \sum_{i=1}^{k} l(\alpha_{i-1}, \alpha_i).$$

Thus the length $l(T)$ is positive if $t_0 < t$ and is negative if $t < t_0$. As was done in §28, we define $\delta(T)$ as the largest of the differences $|t_i - t_{i-1}|$, for $i = 1, 2, \ldots, k$. We will show that the quantity $l(T)$ converges to a limit as $\delta(T) \to 0$. This limit will be called *the length of the curve.* Furthermore, it is equal to

$$\lim_{\delta(T) \to 0} l(T) = \int_{t_0}^{t} + \sqrt{(\varphi'(\tau))^2 + (\psi'(\tau))^2}\, d\tau. \tag{7}$$

Let us prove (7). Choose any two values μ and ν of the parameter t and denote the points $(\varphi(\mu),\, \psi(\mu))$ and $(\varphi(\nu),\, \psi(\nu))$ by β and γ respectively. Write $l(\beta, \gamma)$ for the length of the line segment $[\beta\,\gamma]$. As above, we regard this length as positive if $\mu < \nu$ and negative if $\mu > \nu$. It is clear that

$$l(\beta, \gamma) = \text{sign}(\nu - \mu)\sqrt{[\varphi(\nu) - \varphi(\mu)]^2 + [\psi(\nu) - \psi(\mu)]^2}.$$

Define a function $\rho(\mu, \nu)$ of the interval $[\mu\,\nu]$ by

$$\rho(\mu, \nu) = \frac{l(\beta, \gamma)}{\nu - \mu} - \sqrt{(\varphi'(\nu))^2 + (\psi'(\nu))^2}. \tag{8}$$

We will show that
$$\lim_{|v-\mu|\to 0} \rho(\mu, v)=0. \tag{9}$$

That is, we will prove that the function $\rho(\mu, v)$ is infinitely small (see (3)). Once we have (9), it will be easy to prove (7). Use Lagrange's formula (see §23, (20)) to write

$$\varphi(v)-\varphi(\mu)=(v-\mu)\,\varphi'(\tau_1) \quad \text{and} \quad \psi(v)-\varphi(\mu)=(v-\mu)\,\psi'(\tau_2),$$

where the numbers τ_1 and τ_2 lie in the interval $[\mu\, v]$. We thus have

$$\frac{l(\beta, \gamma)}{v-\mu}=+\sqrt{(\varphi'(\tau_1))^2+(\psi'(\tau_2))^2}.$$

Let us write

$$\omega=(\varphi'(\tau_1))^2+(\psi'(\tau_2))^2 \quad \text{and} \quad w=(\varphi'(v))^2+(\psi'(v))^2.$$

We can rewrite (8) as

$$\rho(\mu, v)=\sqrt{\omega}-\sqrt{w}.$$

The (positive) square root function is uniformly continuous. Therefore, for every positive number ε, there is a positive number ε_1 such that if

$$|\omega-w|<\varepsilon_1,$$

then

$$|\rho(\mu, v)|=|\sqrt{\omega}-\sqrt{w}|<\varepsilon. \tag{10}$$

We also have

$$|\omega-w|\leq|(\varphi'(\tau_1))^2-(\varphi'(v))^2|+|(\psi'(\tau_2))^2-(\psi'(v))^2|.$$

Note too that

$$|v-\tau_1|\leq|v-\mu| \quad \text{and} \quad |v-\tau_2|\leq|v-\mu|.$$

Thus for a given positive number ε_1, we can find a positive number δ such that the inequality

$$|v-\mu|<\delta \tag{11}$$

implies that

$$|(\varphi'(\tau_1))^2-(\varphi'(v))^2|<\tfrac{1}{2}\varepsilon_1 \quad \text{and} \quad |(\psi'(\tau_2))^2-(\psi'(v))^2|<\tfrac{1}{2}\varepsilon_1.$$

That is, for every positive number ε there exists a positive number δ such that (11) implies (10). This proves (9). It follows that

$$l(T)-\sum_{i=1}^{k}(t_i-t_{i-1})\sqrt{(\varphi'(t_i))^2+(\psi'(t_i))^2}$$

$$=\sum_{i=1}^{k}\left[l(\alpha_{i-1}, \alpha_i)-(t_i-t_{i-1})\sqrt{(\varphi'(t_i))^2+(\psi'(t_i))^2}\right]$$

$$=\sum_{i=1}^{k}(t_i-t_{i-1})\rho(t_{i-1}, t_i). \tag{12}$$

In view of (9) and (4), the last expression in (12) has limit zero as $\delta(T) \to 0$. This proves (7). We have therefore proved that the length of a piecewise smooth curve exists and is equal to the integral in (7).

Example. Consider a curve given by the parametric equations (5). It is natural to think of the parameter t as a coordinate for the curve. As we have frequently done already, we will sometimes designate the point on the curve simply by the symbol t. It may happen that two distinct values of t, say t' and t'', yield the same point on the curve. That is, it may happen that

$$(\varphi(t'), \psi(t')) = (\varphi(t''), \psi(t'')).$$

We will think of t' and t'' as distinct points of the curve in spite of this, and call their coincidence in the plane *a self-intersection of the curve*. The length s of the piece of the curve contained between the points t_0 and t is defined by the integral in (7). Plainly this curve length is a function of the parameter t:

$$s = s(t), \tag{13}$$

and we have

$$\frac{ds}{dt} = +\sqrt{(\varphi'(t))^2 + (\psi'(t))^2}. \tag{14}$$

The derivative $\dfrac{ds}{dt}$ is positive everywhere. We can solve the equation (13) for t as a function of s, obtaining the equality

$$t = t(s), \tag{15}$$

with

$$t'(s) = \left(\frac{ds}{dt}\right)^{-1} > 0.$$

I will not give a proof of the fact that the equation (13) can be solved for t as a function of s. Geometrically, it is clear, if we plot a graph of the function $s(t)$ in a coordinate plane. Let us rewrite the parametric equations (5), making use of (15). We get

$$x = \varphi(t(s)) \quad \text{and} \quad y = \psi(t(s)). \tag{16}$$

The equations (16) are parametric equations for the same curve as is defined by (5): we have merely used the new parameter s. The point t_0 corresponds to the value $s = 0$, that is, the origin of the coordinate s for the new parametrization. The part of the curve from the origin to s has curve length s. This parameter s for our curve is extremely useful. To avoid notational complexity, we will suppose that our curve is given parametrically by

$$x = \varphi(s), \quad y = \psi(s).$$

The identity (14) becomes

$$\frac{ds}{ds}=\sqrt{(\varphi'(s))^2+(\psi'(s))^2},$$

which is to say that

$$(\varphi'(s))^2+(\psi'(s))^2=1. \tag{17}$$

Let $\chi(s)$ denote the vector with initial point the origin in our plane and with terminal point $(\varphi(s),\psi(s))$:

$$\chi(s)=(\varphi(s),\psi(s)).$$

Consider the point s as fixed, and let σ be a nearby, movable, point. Draw the straight line containing $\chi(s)$ and $\chi(\sigma)$. Let $\hat{a}$ be the angle that this straight line makes with the positive axis of abscissas in the coordinate plane. It is obvious that

$$\cos\hat{\alpha}=\hat{a}\,\frac{\varphi(\sigma)-\varphi(s)}{\sigma-s}, \qquad \sin\hat{\alpha}=\hat{a}\,\frac{\psi(\sigma)-\psi(s)}{\sigma-s},$$

where $\hat{a}$ is some positive coefficient. We take the limit as $\sigma\to s$ in these equalities; let α be the limiting value of the angle $\hat{\alpha}$, and a the limiting value of the coefficient $\hat{a}$. We obtain the equalities

$$\cos\alpha=a\,\varphi'(s) \quad\text{and}\quad \sin\alpha=a\,\psi'(s).$$

The equality (17) shows that $a=1$, so that

$$\cos\alpha=\varphi'(s) \quad\text{and}\quad \sin\alpha=\psi'(s). \tag{18}$$

Thus the vector

$$\chi'(s)=(\varphi'(s),\psi'(s))$$

has length 1:

$$|\chi'(s)|=1.$$

As $\sigma\to s$, the secant line through $(\varphi(s),\psi(s))$ and $(\varphi(\sigma),\psi(\sigma))$ moves into coincidence with the tangent to our curve at the point s. (In fact, we may *define* the tangent in this fashion.) Thus the vector $\chi'(s)$ is the unit vector tangent to our curve at the point s and headed in the direction of increasing s. The equations (18) show that

$$\tan\alpha=\frac{\psi'(s)}{\varphi'(s)}. \tag{19}$$

Let us differentiate the identity (17) with respect to s. We find that

$$\varphi'(s)\,\varphi''(s)+\psi'(s)\,\psi''(s)=0.$$

Therefore the vectors $\chi'(s)$ and $\chi''(s)$ are perpendicular to each other (see Chapter I, §2, Example 3). Let β be the angle between the vector $\chi''(s)$ and the positive axis of abscissas, and write b for the length $|\chi''(s)|$ of the vector $\chi''(s)$. We have

$$\varphi''(s) = b\cos\beta \quad \text{and} \quad \psi''(s) = b\sin\beta. \tag{20}$$

The angles α and β are functions of the parameter s, although we have not written this explicitly. Let us compute the quantity $\dfrac{d\alpha}{ds}$, which measures the velocity of rotation of the vector χ with respect to the parameter s. This quantity, which is called *the curvature of the curve*, measures how rapidly the curve is changing direction. We differentiate the equality (19) with respect to s:

$$\frac{d\alpha}{ds}\frac{1}{\cos^2\alpha} = \frac{\psi''(s)\,\varphi'(s) - \varphi''(s)\,\psi'(s)}{(\varphi'(s))^2} = \frac{\psi''(s)\,\varphi'(s) - \varphi''(s)\,\psi'(s)}{\cos^2\alpha}.$$

Applying (18) and (20) to this identity, we find that

$$\frac{d\alpha}{ds} = b\sin(\beta - \alpha). \tag{21}$$

Since the vectors $\chi'(s)$ and $\chi''(s)$ are perpendicular to each other, we must have

$$\sin(\beta - \alpha) = \pm 1.$$

Thus (21) implies that

$$\frac{d\alpha}{ds} = \pm b,$$

which is to say that

$$\left|\frac{d\alpha}{ds}\right| = b,$$

the absolute value of the speed of rotation of the vector $\chi(s)$ is equal to b. The quantity $\dfrac{d\alpha}{ds}$ is positive if the rotation of $\chi(s)$ with increasing s is counterclockwise and is negative if this rotation is clockwise. The quantity $b = b(s)$ is *the curvature of the curve at the point s*.

The reciprocal $r = \dfrac{1}{b}$ of the curvature is also studied in differential geometry. This quantity is called the *radius of curvature*. This terminology has the following geometric foundation. On the curve choose three points $s_1 < s_2 < s_3$ that are close together. Through the three points $\chi(s_1)$, $\chi(s_2)$, and $\chi(s_3)$ on the curve, pass a circle, and let $\hat{r}$ be its radius. It turns out that as all three of s_1, s_2, s_3 converge to a common limit s, the radius $\hat{r}$ will converge to the limit $r = \dfrac{1}{b}$. The proof of this interesting geometric fact is not hard, and I recommend that the reader carry it out.

Chapter VII
Analytic Functions

In this chapter we present the fundamentals of the theory of analytic functions, and we carry the theory fairly far. First we define the integral of a complex function of a complex variable on a curve in the complex plane (the plane of the independent variable). We then establish some special cases of Cauchy's integral theorem, which states that the integral of a differentiable complex-valued function of a complex variable is zero when taken over a closed curve. We infer Cauchy's integral formula and give the Laurent series expansion of a function of a complex variable in the neighborhood of an isolated singular point. At the end of the chapter, we analyze the behavior of an analytic function near an essential singularity. We prove that by an appropriate choice of a sequence converging to this singularity, the function can be made to approach any limit we please.

§31. Integration of Functions of a Complex Variable

In §28, we showed how to integrate an arbitrary real-valued continuous function of a real variable. It is natural to ask for the definition of the integral of a function of a complex variable. Here it turns out that continuity does not suffice. We need to suppose that the function to be integrated admits a derivative at every point. With this modest requirement, a function of a complex variable is expansible in a Taylor series at every point. That is, it can be expanded in a power series in $z - z_0$ having positive radius of convergence. Such functions are called *analytic*. Analytic functions play a very important rôle in mathematics. Their study occupies an entire branch of mathematics, and a very large branch indeed. The proof that a differentiable function of a complex variable is analytic is fairly complicated. We will carry it out in several steps, the preparation for which is our next topic.

Open Sets. An analytic function is expansible in a power series with positive radius of convergence about every point z_0 where it is defined. Consequently, the set of points where an analytic function is defined must have the following property. A set G of points in the complex plane is said to be *open* if, for every point z_0 in the set G, there is a positive number r such that all of the points z for which

$$|z - z_0| < r \qquad (1)$$

also belong to the set G. In what follows, we will always suppose that the functions $f(z)$ that we study are defined on open subsets of the complex plane.

We will now find some simple properties of open sets in the complex plane, which will be needed for our study of complex functions.

We consider a bounded closed subset F of the complex plane and a positive number r. Let F_r denote the set of all points ζ in the complex plane such that there is a point z in F such that

$$|z - \zeta| \leq r. \qquad (2)$$

Then the set F_r is closed and bounded. Also, if F is contained in some open set G, there is a positive number r such that F_r is also contained in G.

To prove these two assertions, we consider two sequences of complex numbers,

$$z_1, z_2, \ldots, z_n, \ldots \qquad (3)$$

and

$$\zeta_1, \zeta_2, \ldots, \zeta_n, \ldots, \qquad (4)$$

as well as a bounded sequence

$$r_1, r_2, \ldots, r_n, \ldots \qquad (5)$$

of positive real numbers. The points z_n of the sequence (3) are all to belong to F. We suppose that the points ζ_n of the sequence (4) have the property that

$$|z_n - \zeta_n| \leq r_n \quad \text{for } n = 1, 2, \ldots. \qquad (6)$$

We will first prove that there is a subsequence

$$n_1, n_2, \ldots, n_k, \ldots \qquad (7)$$

of the sequence of all positive integers such that the subsequences of (3) and (4) with the indices (7) both have limits. For the sequence (3), we proved this in §15. So as not to burden our notation, we will suppose that the sequence (3) itself converges to a point z_0 of the set F. The sequence (5) of positive real numbers is bounded. Hence there is a positive number ρ such that $r_n < \frac{1}{2}\rho$ for all indices n. From this and the fact that the sequence (3) converges to z_0, we see that all of the numbers ζ_n of the sequence (4), beginning with some fixed index, satisfy the inequality

$$|\zeta_n - z_0| \leq \rho. \qquad (8)$$

The set of all points ζ such that

$$|\zeta - z_0| \leqq \rho$$

is closed and bounded. Therefore (8) implies that the sequence (4) admits a convergent subsequence. Again, so as not to burden our notation, we will suppose that the sequence (4) itself has a limit, which we write as ζ_0.

To prove that F_r is closed and bounded, we choose any sequence of points (4) in F_r, and we take all of the r_n's in (5) to be equal to r. Then the sequence (3) of points in F exists by the definition of F_r. Choosing a subsequence of the natural numbers as in (7), we see that a subsequence of (4) converges to a point ζ_0, that the corresponding subsequence of (3) converges to a point z_0 of F, and that the inequality

$$|z_0 - \zeta_0| \leqq r$$

holds. Thus F_r is closed. Its boundedness is obvious, since F is bounded.

Now let G be an open set containing F and assume that there is no positive number r for which F_r is contained in G. That is, there is a sequence (5) of positive numbers with limit zero such that every set F_{r_n} contains a point ζ_n not belonging to G. Let z_n be a point of F such that

$$|\zeta_n - z_n| \leqq r_n \quad (n = 1, 2, \ldots). \tag{9}$$

Thus we have sequences (3) and (4) with the positive numbers r_n as in (5) having limit 0. Adjusting our notation, we may suppose that the sequence (3) has limit z_0 in F. Since the numbers r_n have limit zero, condition (6) shows that the limit ζ_0 of the sequence (4) is equal to z_0. Since the point z_0 belongs to F, it also belongs to the open set G. Hence there is a positive number r such that if

$$|\zeta - z_0| < r,$$

then the point ζ also belongs to G. Thus the points of the sequence (4), beginning with some index, must all belong to G. Since they were chosen so as not to belong to G, we have a contradiction.

A particular closed bounded set F that we will use is the set of all points that belong to a curve K with parametric equations

$$x = \varphi(t), \quad y = \psi(t), \tag{10}$$

where $\varphi(t)$ and $\psi(t)$ are piecewise smooth functions on an interval

$$\mu \leqq t \leqq v. \tag{11}$$

Since we are dealing with the complex plane, the points of which are denoted by letters such as z, it is natural to write the pair of equations (10) as the single equation

$$z = z(t) = \varphi(t) + i\psi(t). \tag{12}$$

As in §30, we will consider the set of all points (12), as t runs through the interval (11), as defining a piecewise smooth curve K. Since the interval (11) is closed and bounded and the function $z(t)$ is continuous, the set of points belonging to the curve K is a bounded and closed subset of the complex plane. We regard K as being the path traced out by all of the points (12), in a specified direction, as t runs form μ to ν. The curve K is oriented in this sense. We can also change the parameter t to another parameter $\hat{t}$, the two parameters being connected by the two mutually inverse relations

$$t = t(\hat{t}) \quad \text{and} \quad \hat{t} = \hat{t}(t), \tag{13}$$

so that we have the identities

$$t = t(\hat{t}(t)) \quad \text{and} \quad \hat{t} = \hat{t}(t(\hat{t})).$$

If we replace t in (12) by $t(\hat{t})$, we obtain

$$z = z(t(\hat{t})) = \hat{z}(\hat{t}). \tag{14}$$

This new parametric equation yields the same curve as does (12), provided that the function $t(\hat{t})$ is strictly increasing. (If $t(\hat{t})$ is strictly decreasing, then the curve (14) has orientation opposite to that of (12).) In what follows, we will suppose that the derivative $t'(\hat{t})$ is continuous and never zero, so that it is of constant sign. From this one can prove that also the derivative $\hat{t}(t)$ is continuous and of the same sign as the derivative $t'(\hat{t})$ (see §22, (25)). If the derivatives $t'(\hat{t})$ and $\hat{t}'(t)$ are both positive everywhere, then the curves defined by (12) and (14) have the same orientation, as noted above. If these derivatives are negative, then the two curves have opposite orientations. In this case we denote the curve defined by (14) by the symbol $-K$. This is natural enough, since as t increases, the curve (12) goes in one direction and the curve (14), as $\hat{t}$ increases, goes in the other.

Integration of a Function on a Path. We consider a continuous complex-valued function $f(z)$ defined in a certain open set G of the complex plane. We consider also a curve K as defined above, defined by a parametric equation (12), such that the set F of all points on the curve K lies inside the open set G. We define *the integral of $f(z)$ on the path K*, written as $\int_K f(\zeta)\,d\zeta$, by

$$\int_K f(\zeta)\,d\zeta = \int_\nu^\mu f(z(\tau))\,z'(\tau)\,d\tau. \tag{15}$$

Let us see to what extent the integral of $f(z)$ on K depends upon our choice of parameters. If we consider a new parametrization of K as in (14) that preserves the orientation of K, then the integral (15) does not change. On the other hand, if we reparametrize so as to change the orientation of K, the integral (15) turns into its negative. We write this as

$$\int_{-K} f(\zeta)\,d\zeta = -\int_{K} f(\zeta)\,d\zeta. \tag{16}$$

Thus, when we write the integral (15), K is not merely the object defined by (12) but in fact the geometric path traced out by the moving point $z(t)$ as it traverses the curve K in a specified direction.

Let us prove the assertions of the preceding paragraph. Consider a reparametrization of K as in (13) and (14). We write

$$\hat{\mu} = \hat{t}(\mu) \quad \text{and} \quad \hat{v} = \hat{t}(v).$$

We change the variable of integration in (15) from τ to $\hat{\tau}$, using the relation $\tau = t(\hat{\tau})$. According to the rules given in §25, this yields

$$\int_{K} f(\zeta)\,d\zeta = \int_{\mu}^{v} f(z(\tau))\,z'(\tau)\,d\tau = \int_{\hat{t}(\mu)}^{\hat{t}(v)} f(z(t(\hat{\tau})))\,z'(t(\hat{\tau}))\,t'(\hat{\tau})\,d\hat{\tau}$$

$$= \int_{\hat{\mu}}^{\hat{v}} f(\hat{z}(\hat{\tau}))\,z'(t(\hat{\tau}))\,t'(\hat{\tau})\,d\hat{\tau} = \int_{\hat{\mu}}^{\hat{v}} f(\hat{z}(\hat{\tau}))\,\hat{z}'(\hat{\tau})\,d\hat{\tau}.$$

(See formula (14).) We thus find that

$$\int_{\mu}^{v} f(z(\tau))\,z'(\tau)\,d\tau = \int_{\hat{\mu}}^{\hat{v}} f(\hat{z}(\hat{\tau}))\,\hat{z}'(\hat{\tau})\,d\hat{\tau}. \tag{17}$$

We must now recall that the inequality $\mu < v$ is essential in defining the integral (15): see (11). If the derivative $\hat{t}'(t)$ is positive, we will have $\hat{\mu} < \hat{v}$. Thus the equality (17) shows in this case that two similarly oriented parametrizations of the curve K produce the same value for the integral (15).

However, if the derivative $\hat{t}'(t)$ is negative, then the inequality $\mu < v$ implies that $\hat{\mu} > \hat{v}$. Therefore the right side of (17) does *not* in this case give the integral of $f(z)$ on the (oriented) path, since the lower limit is greater than the upper limit. To identify the right side of (17) in this case, we consider the integral

$$\int_{\hat{v}}^{\hat{\mu}} f(\hat{z}(\hat{\tau}))\,\hat{z}'(\hat{\tau})\,d\hat{\tau}. \tag{18}$$

According to our definition of the integral on a path, (18) is the integral

$$\int_{-K} f(\zeta)\,d\zeta.$$

We also know that

$$\int_{\hat{v}}^{\hat{\mu}} f(\hat{z}(\hat{t}))\,\hat{z}'(\hat{t})\,d\hat{t} = -\int_{\hat{\mu}}^{\hat{v}} f(\hat{z}(\hat{t}))\,\hat{z}'(\hat{t})\,d\hat{t}. \tag{19}$$

(See formula (49) in §28.) Therefore (19) and (17) yield (16). This completes our proof.

Suppose that we have two paths K_1 and K_2 such that the initial point of K_2 is the terminal point of K_1. We form a new curve K by traversing first K_1 and then K_2. We can give K a single parametrization, and formula (46) of §28 yields

$$\int_K f(\zeta)\,d\zeta = \int_{K_1} f(\zeta)\,d\zeta + \int_{K_2} f(\zeta)\,d\zeta. \tag{20}$$

Curve Length as a Parameter. As in §30, Example, we can take the length of the curve s from the point $z(0)$ to the point $z(s)$ as our parameter t. This gives us

$$|z'(s)| = 1. \tag{21}$$

This follows from formula (17) in §30:

$$|z'(s)| = |\varphi'(s) + i\psi'(s)| = +\sqrt{(\varphi'(s))^2 + (\psi'(s))^2} = 1.$$

The identity (21) yields a useful inequality. Suppose that we have

$$|f(z(s))| \leqq m$$

for all values of s in our parameter interval. From the inequality (58) in §28, we find that

$$\left|\int_K f(\zeta)\,d\zeta\right| = \left|\int_{\mu}^{v} f(z(s))\,z'(s)\,ds\right| \leqq m(v - \mu), \tag{22}$$

the number $v - \mu$ being the length of the curve K.

A Comparison of the Integral Over K with the Integral Over a Broken Line L Inscribed in K. We break up our path of integration K into pieces by introducing a subdivision of the parameter interval $[\mu v]$:

$$T = (t_0, t_1, \ldots, t_j, \ldots, t_k) \quad \text{with} \quad t_0 = \mu \text{ and } t_k = v.$$

We join the point $z(t_{j-1})$ with the point $z(t_j)$ by a line segment, which we will call L_j ($j = 1, 2, \ldots, k$). Let $L = L(T)$ be the path obtained by traversing first L_1 from $z(t_0)$ to $z(t_1)$, then L_2 from $z(t_1)$ to $z(t_2)$, ..., and finally L_k from $z(t_{k-1})$ to $z(t_k)$. We say that the broken line $L(T)$ is *inscribed in the curve K*. Our task is to compare the integral of a function $f(z)$ over K with its integrals over the broken line $L(T)$. We will prove in fact that

$$\int_K f(\zeta) \, d\zeta = \lim_{\delta(T) \to 0} \int_{L(T)} f(\zeta) \, d\zeta. \qquad (23)$$

That is, we can approximate the integral over K just as closely as we please by integrals over the broken lines $L(T)$. This fact will play a vital rôle in our later analysis. We parametrize the line segments L_j in a natural way, so that the point $z^*(t)$ traverses L_j in the correct direction as t runs from t_{j-1} to t_j. We do this by defining

$$z^*(t) = \frac{z(t_{j-1})(t_j - t) + z(t_j)(t - t_{j-1})}{t_j - t_{j-1}} \quad \text{for } t_{j-1} \leq t \leq t_j. \qquad (24)$$

It is immediate that $z^*(t)$ traverses the line segment L_j as t runs from t_{j-1} to t_j. As t runs from μ to v, the point $z^*(t)$ traces out the broken line $L(T)$ in the correct direction. We take this parametrization for the line $L(T)$.

Since the derivative $z'(t)$ appears in the definition of the integral (15), we must compute $z^{*\prime}(t)$. On the interval $t_{j-1} < t < t_j$, (24) shows that

$$z^{*\prime}(t) = \frac{z(t_j) - z(t_{j-1})}{t_j - t_{j-1}} = \frac{\varphi(t_j) - \varphi(t_{j-1})}{t_j - t_{j-1}} + i \frac{\psi(t_j) - \psi(t_{j-1})}{t_j - t_{j-1}}.$$

Lagrange's formula (see §23, (20)) shows that

$$\frac{\varphi(t_j) - \varphi(t_{j-1})}{t_j - t_{j-1}} = \varphi'(\alpha_j)$$

and

$$\frac{\psi(t_j) - \psi(t_{j-1})}{t_j - t_{j-1}} = \psi'(\beta_j),$$

where the numbers α_j and β_j belong to the closed interval $[t_{j-1} t_j]$. Thus for all t in the interval $[t_{j-1} t_j]$, we have

$$z'(t) - z^{*\prime}(t) = [\varphi'(t) - \varphi'(\alpha_j)] + i[\psi'(t) - \psi'(\beta_j)].$$

The function $z'(t)$ is piecewise continuous and so is piecewise uniformly continuous. It follows that for all t in the entire parameter interval $[\mu v]$, we have

$$|z'(t) - z^{*\prime}(t)| \leq \rho(T), \qquad (25)$$

where

$$\lim_{\delta(T) \to 0} \rho(T) = 0.$$

Here $\rho(T)$ is a nonnegative real number that depends upon the subdivision T (and of course on the curve K, which is fixed). We also have

$$z(t)-z^*(t)=\int_\mu^t [z'(\tau)-z^{*\prime}(\tau)]\,d\tau.$$

Using formula (58) in §28, we find that

$$|z(t)-z^*(t)|\leq\rho(T)(v-\mu). \tag{26}$$

As before, we write F for the set of all points belonging to the curve K. Choose a positive number r so small that the set F_r is contained in the open set G. (Recall that G is an open set in the complex plane, where the integrand $f(z)$ is defined, that contains all of the points of the curve K.) We suppose that all of the paths $L(T)$ under consideration lie entirely within the set F_r. (We can secure this by considering only T's for which $\delta(T)$ is sufficiently small.) The function $f(z)$ is uniformly continuous on the closed bounded set F_r. Accordingly, the inequality (26) shows that

$$|f(z(t))-f(z^*(t))|<\rho_1(T), \tag{27}$$

where $\rho_1(T)$ is a nonnegative number depending upon the subdivision T for which

$$\lim_{\delta(T)\to 0}\rho_1(T)=0.$$

We infer from (25) and (27) that

$$|f(z(t))\,z'(t)-f(z^*(t))\,z^{*\prime}(t)|\leq\rho_2(T), \tag{28}$$

where $\rho_2(T)$ is a nonnegative number depending upon the subdivision T for which

$$\lim_{\delta(T)\to 0}\rho_2(T)=0.$$

To see this, make the following estimates:

$$\begin{aligned}
&|f(z(t))\,z'(t)-f(z^*(t))\,z^{*\prime}(t)|\\
&=|[f(z(t))\,z'(t)-f(z(t))\,z^{*\prime}(t)]-[f(z^*(t))\,z^{*\prime}(t)-f(z(t))\,z^{*\prime}(t)]|\\
&\leq|f(z(t))[z'(t)-z^{*\prime}(t)]|+|[f(z^*(t))-f(z(t))]\,z^{*\prime}(t)|\\
&\leq m\,\rho(T)+\rho_1(T)\cdot m,
\end{aligned}$$

where m is a positive number that exceeds both $|f(z(t))|$ and $|z^{*\prime}(t)|$ for all t in the parameter interval $[\mu v]$. Such a number m exists because $f(z(t))$ is continuous and hence bounded, and $z^{*\prime}(t)$ assumes only a finite number of values, as we have already shown. Therefore we can define

$$\rho_2(T) = m(\rho(T) + \rho_1(T))$$

and so obtain (28).

We now use (28) and the estimate (58) of §28 to write

$$\left| \int_K f(\zeta)\,d\zeta - \int_L f(\zeta)\,d\zeta \right| = \left| \int_\mu^\nu [f(z(\tau))\,z'(\tau) - f(z^*(\tau))\,z^{*\prime}(\tau)]\,d\tau \right|$$

$$\leq \rho_2(T)(\nu - \mu). \tag{29}$$

The relations (29) and $\lim_{\delta(T) \to 0} \rho_2(T) = 0$ immediately give (23), which we wished to prove.

Example. In proving (23), I have twice used in implicit form the notion of uniform convergence, which we will have to apply explicitly later on. We pause here to define the concept explicitly.

Consider a sequence

$$f_1(z), f_2(z), \dots, f_n(z), \dots \tag{30}$$

of real- or complex-valued functions of a real or complex variable, as well as a fixed function $f(z)$, all defined on some fixed set of real or complex numbers. Suppose that for every positive number ε, there exists a positive integer ν such that the inequality

$$|f_n(z) - f(z)| < \varepsilon$$

holds for all points z where the functions are defined, provided that n be greater than ν. Then we say that the sequence of functions $f_n(z)$ *converges uniformly to the limit function* $f(z)$. Suppose that the variable z runs through an interval $[\mu\sigma]$ in the real number system and that the functions $f_n(t)$ and $f(t)$ are piecewise continuous. Then if $f_n(t)$ converges uniformly to $f(t)$, it is easy to see that

$$\lim_{n \to \infty} \int_\mu^t f_n(\tau)\,d\tau = \int_\mu^t f(\tau)\,d\tau.$$

To prove this, we have only to write

$$\left| \int_\mu^t f_n(\tau)\,d\tau - \int_\mu^t f(\tau)\,d\tau \right| = \left| \int_\mu^t (f_n(\tau) - f(\tau))\,d\tau \right| \leq |t - \mu|\,\varepsilon.$$

We used estimates of this sort in proving (26) and (29), using instead of the index n a variable subdivision T of our parameter interval.

We described early on the connection between sequences and series (§16, (2)). Using this connection, it is simple to define the notion of a uniformly convergent series of functions and to prove that a uniformly convergent series of functions can be integrated term by term.

§ 32. Cauchy's Theorem

Cauchy's theorem is the central theorem of the theory of analytic functions. In the example at the end of this section, we will state this theorem in its general form and give a proof, which is not excessively pedantic. In the body of this section, we limit ourselves to some particularly important special cases of the theorem.

We will use curves K in the complex plane, as these were defined in § 31. We call a curve *closed* if its initial point and its terminal point coincide:

$$z(\mu) = z(v), \tag{1}$$

the curve itself being given by

$$z = z(t) = \varphi(t) + i\psi(t), \tag{2}$$

where the parameter t runs through the interval $[\mu v]$.

Consider a differentiable function $f(z)$ and its integral over a closed curve K:

$$\int_K f(\zeta) \, d\zeta,$$

the integral being defined as in § 31, (15). Cauchy's theorem is concerned with such integrals. We first present some preliminaries.

Suppose that we have a complex-valued function $f(z)$ defined on a certain open set G in the complex plane. Suppose that $f(z)$ admits an indefinite integral $h(z)$:

$$h'(z) = f(z), \tag{3}$$

where the function $h(z)$ is defined on the same open set G as $f(z)$ and is single-valued. (We must underline this restriction because we have already had occasion to consider multiple-valued functions that are primitives, for example $\ln z$. Of course we ordinarily mean the term "function" to refer to a single-valued function.) If (3) holds, we can easily compute the integral of $f(z)$ on a path K:

$$\int_K f(\zeta) \, d\zeta = h(z(v)) - h(z(\mu)). \tag{4}$$

Let us prove (4). The rule for differentiating a composite function shows that

$$\frac{dh(z(t))}{dt} = h'(z(t)) \, z'(t) = f(z(t)) \, z'(t). \tag{5}$$

Applying § 25, formulas (20) and (21), we find that

$$\int_K f(\zeta) \, d\zeta = \int_\mu^v f(z(\tau)) \, z'(\tau) \, d\tau = h(z(v)) - h(z(\mu)),$$

which is (4).

In particular, if K is a closed curve, (4) shows that

$$\int_K f(\zeta)\,d\zeta = 0.$$

We will now prove a special case of Cauchy's theorem, which in fact contains the heart of the theorem.

Cauchy's Theorem for a Triangle. Suppose that we have a triangle in our open set G with vertices a, b, c. That is, we suppose that the boundary T_0 of the triangle with these vertices lies entirely within G and also that the region $\hat{T}_0$ consisting of T_0 together with all of the points that lie inside T_0 lies entirely within G. Let T_0 denote not only the boundary of the triangle but also the path that consists of the line segment $[ab]$ followed by the line segment $[bc]$ followed by the line segment $[ca]$. We write the initial point of each line segment first and its terminal point last. We will suppose that our notation is chosen so that the path T_0 is traced out counterclockwise.

Cauchy's theorem for a triangle states that if $f(z)$ is a complex function defined in G that has a derivative at every point of G, then the integral of $f(z)$ around T_0 is zero:

$$\int_{T_0} f(\zeta)\,d\zeta = 0. \tag{6}$$

We will prove (6) by contradiction. We assume that

$$\left| \int_{T_0} f(\zeta)\,d\zeta \right| \geq \alpha > 0. \tag{7}$$

Let c_1, a_1, b_1 be the midpoints of the line segments $[ab]$, $[bc]$, $[ca]$, respectively. The line segments $[a_1 b_1]$, $[b_1 c_1]$, and $[c_1 a_1]$ divide the original triangular region $\hat{T}_0$ into four triangular regions. We denote their boundaries, oriented in the counterclockwise direction, by

$$T_1^1, T_1^2, T_1^3, T_1^4. \tag{8}$$

Each of the triangles (8) is similar to the original triangle T_0 with coefficient of similarity equal to $\frac{1}{2}$. Observe next that

$$\int_{T_0} f(\zeta)\,d\zeta = \sum_{j=1}^{4} \int_{T_1^j} f(\zeta)\,d\zeta. \tag{9}$$

This equality holds because the inner edges $[a_1 b_1]$, $[b_1 c_1]$, and $[c_1 a_1]$ are all described twice in the integrals on the right side of (9), once in each direction, so that their contributions cancel each other (formula (16) in §31). The outer edges are all described exactly once, in the counterclockwise direction. (The reader may wish to make a sketch to illustrate this argument.)

We now look at (7) and (9). There must be at least one triangle T_1^j, which we designate as T_1, for which

$$|\int_{T_1} f(\zeta)\,d\zeta| \geq \tfrac{1}{4}\alpha.$$

We divide the triangle T_1 into four subtriangles just as we did with T_0. The same argument shows that there is at least one of these yet smaller triangles, say T_2, with the property that

$$|\int_{T_2} f(\zeta)\,d\zeta| \geq (\tfrac{1}{4})^2\,\alpha.$$

The triangle T_2 is similar to the triangle T_0 with coefficient of similarity $(\tfrac{1}{2})^2$. We continue this construction for every positive integer. We obtain a nested sequence of triangles

$$T_0, T_1, \ldots, T_n, \ldots,$$

where T_n is similar to T_0 with coefficient of similarity $(\tfrac{1}{2})^n$ and the property that

$$|\int_{T_n} f(\zeta)\,d\zeta| \geq (\tfrac{1}{4})^n\alpha. \tag{10}$$

Let $\hat{T}_n$ denote the triangular region consisting of the triangle T_n together with all of the points that lie in the interior of T_n. It is clear that $\hat{T}_{n+1}$ is contained in $\hat{T}_n$ for $n=0, 1, 2, \ldots$. We will prove that there is a point $\hat{z}$ that belongs to all of the triangles

$$\hat{T}_0, \hat{T}_1, \ldots, \hat{T}_n, \ldots. \tag{11}$$

To prove this, let l be the length of the longest of the sides of the original triangle T_0. The length of the longest of the sides of the triangle T_n is obviously $(\tfrac{1}{2})^n\,l$. Therefore the distance between any two points of the triangular region $\hat{T}_n$ does not exceed $(\tfrac{1}{2})^n\,l$. Now choose a point z_n in the triangular region $\hat{T}_n$, for $n=0, 1, 2, \ldots$. We will prove that the sequence

$$z_0, z_1, \ldots, z_n, \ldots \tag{12}$$

is a Cauchy sequence and that its limit $\hat{z}$ belongs to all of the triangular regions $\hat{T}_n$. Suppose that v is any natural number and that p and q are any natural numbers greater than v. Then the points z_p and z_q belong to $\hat{T}_v$ and so the distance between z_p and z_q does not exceed $(\tfrac{1}{2})^v\,l$. Therefore the sequence (12) is a Cauchy sequence.

All of the points of the sequence (12) beginning with z_v belong to the triangular region $\hat{T}_v$. Therefore the limit $\hat{z}$ belongs to $\hat{T}_v$, since $\hat{T}_v$ is a closed set. Therefore $\hat{z}$ belongs to all of the triangular regions listed in (11).

The function $f(z)$ being differentiable everywhere in G, it is differentiable at the point $\hat{z}$. That is to say, the limit

$$\lim_{z \to \hat{z}} \frac{f(z) - f(\hat{z})}{z - \hat{z}} = f'(\hat{z})$$

exists. We may write this relation in the form

$$\frac{f(z) - f(\hat{z})}{z - \hat{z}} - f'(\hat{z}) = \gamma(z), \tag{13}$$

where $|\gamma(z)| \leq \gamma$, and γ is a number for which $\lim_{z \to \hat{z}} \gamma = 0$.

We rewrite the equality (13) as

$$f(z) = f(\hat{z}) + f'(\hat{z})(z - \hat{z}) + \gamma(z)(z - \hat{z}). \tag{14}$$

We now integrate $f(z)$, written in the form (14), around the path T_n. The sum of the first two terms on the right side of (14) has integral 0 around the path T_n. To see this, observe that

$$\frac{\partial}{\partial z} [f(\hat{z})(z - \hat{z}) + \tfrac{1}{2} f'(z - \hat{z})^2] = f(\hat{z}) + f'(\hat{z})(z - \hat{z}).$$

That is, the sum of the first two terms on the right side of (14) has a primitive, and so formula (4) shows that the integral of these terms around T_n is 0. We will now estimate the integral of the third summand

$$\gamma(z)(z - \hat{z})$$

around T_n, using the inequality (22) of §31. Note first that the length of the path of integration does not exceed $(\tfrac{1}{2})^n 3 l$. Since the points z and $\hat{z}$ belong to the region $\hat{T}_n$, we have

$$|z - \hat{z}| \leq (\tfrac{1}{2})^n l.$$

It follows that

$$|\int_{T_n} \gamma(z)(z - \hat{z}) \, dz| \leq (\tfrac{1}{4})^n 3 l^2 \gamma,$$

where

$$\lim_{n \to \infty} \gamma = 0. \tag{15}$$

Combining (15) with the above calculation of the integral of the first two terms in (14), we find that

$$|\int_{T_n} f(\zeta) \, d\zeta| \leq (\tfrac{1}{4})^n 3 l^2 \gamma. \tag{16}$$

Since $\gamma \to 0$ as $n \to \infty$, the estimates (16) and (10) contradict each other. Therefore (6) is proved.

Observe that our proof of (6) does not require that $f'(z)$ be continuous, only that it exist.

We now draw two inferences from the equality (6): both are also special cases of Cauchy's theorem.

Corollary 1. Again we consider an open set G in the complex plane and a differentiable function $f(z)$ defined in G. Let K be a closed curve lying entirely within G, defined parametrically by the equation (2) and subject to the condition (1). Choose a point a in G and suppose that the line segments $[az(t)]$, as $z(t)$ traces out the curve K, all lie entirely within G. We then have

$$\int_K f(\zeta)\,d\zeta = 0. \qquad (17)$$

To prove (17), we begin by inscribing a broken line in the curve K, defined as in §31 by a subdivision

$$T = (t_0, t_1, \ldots, t_j, \ldots, t_k) \quad \text{with} \quad t_0 = \mu \text{ and } t_k = \nu,$$

of the parameter interval $[\mu\nu]$. If the subdivision T is sufficiently fine, that is, if $\delta(T)$ is sufficiently small, then the entire broken line $L(T)$ will be very close to the curve K. From this it follows, since G is an open set, that all of the line segments $[a z^*(t)]$, going from the point a in G to a point $z^*(t)$ on the broken line $L(T)$, lie entirely within G. Then consider a triangular path K_j formed by the triangle with sides

$$[az(t_{j-1})], \quad [z(t_{j-1})z(t_j)], \quad \text{and} \quad [z(t_j)a].$$

This triangular path bounds a triangular region $\hat{K}_j$, which under our hypothesis on G and for sufficiently fine T, lies entirely within G. Therefore (6) holds for every triangular path K_j:

$$\int_{K_j} f(\zeta)\,d\zeta = 0. \qquad (18)$$

Add the identities (18) for all $j = 1, 2, \ldots, k$ and obtain

$$\sum_{j=1}^{k} \int_{K_j} f(\zeta)\,d\zeta = 0. \qquad (19)$$

Note that we traverse all of the triangular paths K_j in the counterclockwise direction. Now look at the oriented line segments $[a z(t_j)]$ and $[z(t_j)a]$ $(j = 0, 1, \ldots, k)$. Each of them occurs just once in the oriented triangles K_j. (Observe that $z(t_0) = z(t_k)$, and make a sketch to illustrate the geometry involved.) Hence the integrals over these pairs of line segments cancel each other, and so the left side of (19) is equal to $\int_{L(T)} f(\zeta)\,d\zeta$. We have therefore proved that

$$\int_{L(T)} f(\zeta)\,d\zeta = 0.$$

We proved in § 31, (23), that

$$\int_K f(\zeta)\,d\zeta = \lim_{\delta(T)\to 0} \int_{L(T)} f(\zeta)\,d\zeta.$$

Combine the last two equalities to obtain (17).

Consider a closed curve defined by (2) and (1). We can obviously traverse this curve beginning anywhere on it without changing the value of the integral of a function on it. That is, when we integrate a function over a closed curve, we need not specify the initial point as in (1), but only the curve itself.

A special case of Corollary 1 is provided by a circle K such that the circle itself and all of the points in its interior lie in our open set G. We can then take a in Corollary 1 to be the center of the circle. Thus we find that

$$\int_K f(\zeta)\,d\zeta = 0. \tag{20}$$

Corollary 2. Let G be an open set in the complex plane and $f(z)$ a differentiable function defined throughout G. Let K_1 and K_2 be concentric circles with common center o_2, with the radius of K_1 larger than the radius of K_2. Consider the annular region contained between the circles K_1 and K_2. Suppose that there is a circle K_3 with center o_3 such that K_3 and all of the points in its interior lie in the annular region just specified. Suppose finally that all of the points on the circle K_1 or in its interior belong to the open set G, with the possible exception of the two points o_2 and o_3. We take the circles K_1, K_2, and K_3 as paths of integration for our function $f(z)$, all three described in the counterclockwise direction. We then have

$$\int_{K_1} f(\zeta)\,d\zeta = \int_{K_2} f(\zeta)\,d\zeta + \int_{K_3} f(\zeta)\,d\zeta. \tag{21}$$

To prove (21), we will make use of Fig. 68. To render our verbal description simple, we will suppose that the line P going through o_2 and o_3 is horizontal and that o_2 lies to the left of o_3. This assumption has nothing whatever to do with the position of the coordinate axes in the complex plane. We have simply, for the moment, rotated our drawing to make our description easy. The line P divides each of the circles K_j into an upper half-circle K'_j and a lower half-circle K''_j. Each of the circles K_j intersects the line P twice: let b_j be the intersection on the right and c_j the intersection on the left. Let R denote the subset of G consisting of all points that are (strictly) interior to K_1 and (strictly) exterior to both K_2 and K_3. The line P

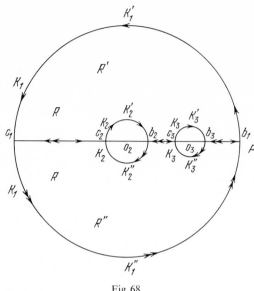

Fig. 68

divides the open set R into an upper half R' and a lower half R''. Let C' be the closed path that is the boundary of R' traversed in the counterclockwise direction. Let C'', in like manner, be the closed path that is the boundary of R'', also traversed in the counterclockwise direction. Let a' be any point in the open set R' and a'' any point in the open set R''. It is easy to see (see Fig. 68 once again and recall that all points interior to K_1 except possibly o_2 and o_3 belong to G) that the entire line segment starting from a' and going to any point on C' lies within G. A similar statement holds of course for the symmetric region R'', the point a'', and the boundary curve C''. Corollary 1 shows accordingly that

$$\int_{C'} f(\zeta)\,d\zeta = 0 \tag{22}$$

and

$$\int_{C''} f(\zeta)\,d\zeta = 0. \tag{23}$$

The closed path C' consists of six pieces, three semicircles and three line segments. We write this path as follows, paying attention to the fact that it is traversed counterclockwise:

$$C' = (K_1', [c_1 c_2], -K_2', [b_2 c_3], -K_3', [b_3 b_1]).$$

In like manner, we write C'':

$$C'' = (K_1'', [b_1 b_3], -K_3'', [c_3 b_2], -K_2'', [c_2 c_1]).$$

Observe now that the line segments appearing in C' and C'' are the negatives of each other, that the semicircles K'_1 and K''_1 are traversed counterclockwise, and that the semicircles K'_j and K''_j ($j=2,3$) are traversed clockwise. Add (22) and (23), taking account of formulas (16) and (20) of §31. We find that

$$\int_{K_1} f(\zeta)\,d\zeta + \int_{-K_2} f(\zeta)\,d\zeta + \int_{-K_3} f(\zeta)\,d\zeta = 0,$$

and therefore (21) holds.

Formally each of the three integrals in (21) depends upon the radius r_j of the circle K_j. However, (21) holds for *all* radii r_1, r_2, r_3 subject only to the inequalities that preserve the geometry sketched in Fig. 68. It follows that each of the integrals in (21) is in fact independent of the corresponding radius.

Cauchy's Integral Formula. We will now make use of (21) to obtain a famous formula of Cauchy. Write the origin o_2 of K_1 and K_2 as u (a complex number, of course). We consider the function

$$g(z)=(z-u)^n$$

for an arbitrary integer n, positive, negative, or zero. We wish to compute the integral

$$h(u)=\int_{K_2} (\zeta-u)^n\,d\zeta.$$

We parametrize the circle K_2 by setting

$$z(t)=u+r_2\,e^{it}.$$

As t goes from 0 to 2π, the point $z(t)$ traverses the circle K_2 exactly once in the counterclockwise direction. We also have

$$z'(t)=ir_2\,e^{it}.$$

Formula (15) of §31 shows that

$$h(u)=\int_0^{2\pi} r_2^n e^{int} ir_2\,e^{it}\,d\tau = ir_2^{n+1}\int_0^{2\pi} e^{i(n+1)\tau}\,d\tau.$$

If $n+1\neq0$, the integrand in the last integral of this identity has a primitive $\dfrac{1}{i(n+1)}e^{i(n+1)\tau}$. From §32, formula (4), we see that $h(u)=0$, or

$$\int_{K_2} (\zeta-u)^n\,d\zeta=0 \qquad\qquad (24)$$

if $n \neq -1$. For $n = -1$, we get

$$h(u) = r_2^0 i \int_0^{2\pi} d\tau = 2\pi i.$$

That is to say, we have

$$\int_{K_2} \frac{d\zeta}{\zeta - u} = 2\pi i. \tag{25}$$

The last identity admits a generalization of extreme importance. Let $f(z)$ be a function that is differentiable in an open set that contains the entire closed disk bounded by K_1. We write

$$g(z) = \frac{f(z)}{z - u},$$

and then compute the integral

$$h(u) = \int_{K_2} \frac{f(\zeta)}{\zeta - u} d\zeta.$$

We have

$$g(z) = \frac{f(u)}{z - u} + \frac{f(z) - f(u)}{z - u} = \frac{f(u)}{z - u} + f'(u) + \gamma(z), \tag{26}$$

where $\gamma(z)$ is a function such that $|\gamma(z)| < \gamma$ and, for z belonging to K_2, we have $\lim_{r_2 \to 0} \gamma = 0$. The integral over the circle K_2 of the function on the left side of (26) does not depend upon r_2. The same is true of the first two summands after the second equality sign in (26). We also have

$$|\int_{K_2} \gamma(\zeta) d\zeta| \leq \gamma \cdot 2\pi r.$$

It follows that

$$\lim_{r_2 \to 0} \int_{K_2} \gamma(\zeta) d\zeta = 0. \tag{27}$$

The integral on the left side of (27) does not depend upon r_2, as the foregoing discussion proves, and from (27) we infer that this integral is equal to 0. Going back to (26), therefore, and taking account of (25), we find that

$$\int_{K_2} \frac{f(\zeta)}{\zeta - u} d\zeta = f(u) \int_{K_2} \frac{d\zeta}{\zeta - u} + f'(u) \int_{K_2} d\zeta = 2\pi i f(u).$$

We write this identity anew, denoting by K a circle with center u traversed in the counterclockwise direction:

$$\int_K \frac{f(\zeta)}{\zeta - u} d\zeta = 2\pi i f(u). \tag{28}$$

In writing (28), we suppose that the circle K and its entire interior are contained in an open set G at every point of which $f(z)$ is differentiable. The identity (28) is *Cauchy's integral formula*. It is the most important single formula in the entire theory of analytic functions.

Example. We now present a general formulation of Cauchy's theorem and its proof, which is not totally pedantic. We first consider an open subset G of the complex plane and a function $f(z)$ defined and differentiable throughout G. We consider a set $K_0, K_1, \ldots, K_m$ of closed, piecewise smooth paths all lying in G. We suppose that all of them are oriented counterclockwise. We suppose also that no two of the paths intersect each other and that none of them has a self-intersection. We suppose that all of the closed curves

$$K_1, K_2, \ldots, K_m \tag{29}$$

lie in the interior of the the path K_0 and that none of the curves (29) lies in the interior of any of the others. Let R denote the part of the complex plane that is in the interior of K_0 and is exterior to all of the curves (29). We will suppose that the set R is contained in the open set G. We then have

$$\int_{K_0} f(\zeta)\, d\zeta = \sum_{j=1}^{m} \int_{K_j} f(\zeta)\, d\zeta. \tag{30}$$

To prove (30), we first divide up the entire complex plane into tiny squares, by means of horizontal and vertical lines. We will suppose that these squares are so small that each square that intersects the set R is contained entirely within G. Let

$$C_1, C_2, \ldots, C_n$$

be the set of all of the squares that intersect R. (We think of the square as a plane region, not merely the boundary of the square.) Let B_k denote the intersection of C_k with R, and let A_k denote the boundary of the set B_k. If the square C_k lies entirely within R, then A_k is simply the boundary of the square C_k. If C_k is not wholly contained in R, then A_k consists of some line segments, which enter into the boundary of C_k, and also of some pieces of the curves that bound the set R. Each of these pieces of curve belong to one of the curves K_j $(j = 0, 1, \ldots, m)$. Within each set B_k, choose an arbitrary point a_k and join it by a line segment to each point of its boundary A_k. All of these line segments plainly belong to the open set G. We therefore have

$$\int_{A_k} f(\zeta)\, d\zeta = 0. \tag{31}$$

We take the closed curve A_k to be oriented in the counterclockwise direction. Every line segment of the boundary of A_k that belongs to the square C_k

also belongs to another square C_l and appears in the boundary curve of C_l in the opposite direction. Every other part of the boundary A_k belongs to one of the curves K_j, with $j=0, 1, \ldots, m$. If it belongs to K_0, it is traversed in the same direction as in K_0. If it belongs to one of the curves (29), then it is traversed in the opposite direction in A_k to its direction in K_j. Adding the identities (31) from 1 to n, we find that

$$\int_{K_0} f(\zeta)\,d\zeta + \sum_{j=1}^{m} \int_{-K_j} f(\zeta)\,d\zeta = 0.$$

This identity gives us (30).

Our proof is formally not irreproachable. For example, we take it as evident that a closed curve without self-intersections divides the plane into two parts, its interior and its exterior. This fact is by no means easy to prove, but here we take it as evident. A second flaw is that some of the curves A_j may not be nearly so simple as they appear at first glance. For example, they can contain an infinite number of curved pieces.

§33. Taylor Series and Laurent Series

Let G be an open subset of the complex plane, and let $f(z)$ be a function that is defined and differentiable throughout G. We will prove that for every point z_0 of G, there is a Taylor series expansion for the function $f(z)$ with positive radius of convergence:

$$f(z) = a_0 + a_1(z-z_0) + a_2(z-z_0)^2 + \cdots + a_n(z-z_0)^n + \cdots. \qquad (1)$$

(See §26.) We will find a lower bound for the radius of convergence of the series (1). We will prove in fact that if the open disk consisting of all z such that

$$|z-z_0| < r'$$

is contained in G, then the radius of convergence r of (1) is no smaller than r':

$$r \geqq r'.$$

We will also consider the case in which z_0 does not belong to G but for some positive number r', the set of all z such that

$$0 < |z-z_0| < r'$$

is contained in G. In this case, we say that the function $f(z)$ has *an isolated singularity at the point* z_0. The function $f(z)$ is not defined at z_0 but is defined and differentiable at all other points of some open disk with center at z_0. In this case, the function $f(z)$ is expansible in what is called a *Laurent series*. That is to say, $f(z)$ is the sum of two functions

$$f(z)=f_+(z)+f_-(z), \tag{2}$$

where the function $f_+(z)$ admits a power series expansion of the form (1) in nonnegative powers of the variable $z-z_0$, and the function $f_-(z)$ admits a power series expansion in negative powers of $z-z_0$:

$$f_-(z)=\frac{a_{-1}}{z-z_0}+\frac{a_{-2}}{(z-z_0)^2}+\cdots+\frac{a_{-n}}{(z-z_0)^n}+\cdots. \tag{3}$$

The series (3) converges for all

$$z\neq z_0. \tag{4}$$

Let us first give an example of a function that we know how to expand in a Laurent series. This is the function

$$f(z)=e^{1/z}, \tag{5}$$

defined for all complex numbers $z\neq 0$, and admitting the expansion

$$e^{1/z}=1+\frac{1}{z}+\frac{1}{2!}\frac{1}{z^2}+\cdots+\frac{1}{n!}\frac{1}{z^n}+\cdots. \tag{6}$$

(See §19, formula (4).) Thus for the function (5) we have $f_+(z)=1$, and $f_-(z)$ is the sum of all of the terms in (6) with negative powers of z.

Plainly the Taylor series expansion (1) is a special case of the Laurent series expansion (2). This occurs when $f(z_0)$ can be defined in such a way that $f(z)$ is differentiable in the entire open disk consisting of the points z at distance less than r' from z_0. Then the function $f_-(z)$ is simply taken to be identically zero.

Let us prove the validity of the Laurent series expansion (2). Consider a circle K_1 with center at the point z_0 such that the circle itself and all of the points interior to it, except possibly z_0, belong to the open set G. Let w be a point in the interior of K_1 that is different from z_0. Choose a circle K_2 with center z_0 and a circle K_3 with center w such that both K_2 and K_3, with their interiors, lie in the interior of K_1 and so that each of K_2 and K_3 lies entirely in the exterior of the other. We now write

$$g(z)=\frac{f(z)}{z-w}.$$

We will apply formula (21) of §32 to this function $g(z)$, taking z_0 for the point o_2 and w for the point o_3. For notational convenience, we will write each of the three integrals occurring in (21) of §32 with a special notation:

$$h_m(w) = \frac{1}{2\pi i} \int_{K_m} \frac{f(\zeta)}{\zeta - w} d\zeta \qquad (m = 1, 2, 3).$$

Formula (21) of §32 is then written as

$$h_3(w) = h_1(w) - h_2(w). \tag{7}$$

(Recall that each of the circles K_m is traversed in the counterclockwise direction.) We now compute the functions $h_3(w)$, $h_1(w)$, and $h_2(w)$. Cauchy's integral formula (formula (28) in §32) shows at once that

$$h_3(w) = f(w). \tag{8}$$

To compute $h_1(w)$ we rewrite the integrand in the form

$$\frac{f(\zeta)}{\zeta - w} = \frac{f(\zeta)}{(\zeta - z_0) - (w - z_0)} = \frac{f(\zeta)}{\zeta - z_0} \cdot \frac{1}{1 - \dfrac{w - z_0}{\zeta - z_0}}$$

$$= \sum_{n=0}^{\infty} f(\zeta) \frac{(w - z_0)^n}{(\zeta - z_0)^{n+1}}. \tag{9}$$

In writing (9), we expand the expression

$$\frac{1}{1 - \dfrac{w - z_0}{\zeta - z_0}}$$

in the form of an (infinite) geometric progression. This can be done, since

$$\left| \frac{w - z_0}{\zeta - z_0} \right| = \rho < 1.$$

(Note that the point ζ lies on the circle K_1 for all values of the integrand for $h_1(w)$, while w lies in the interior of this circle.)

The series in the second line of (9) converges uniformly in the variable of integration ζ (see the Example in §31). Accordingly this series can be integrated term by term. We find that

$$h_1(w) = \frac{1}{2\pi i} \sum_{n=0}^{\infty} (w - z_0)^n \int_{K_1} \frac{f(\zeta) d\zeta}{(\zeta - z_0)^{n+1}}.$$

It follows that

$$h_1(w) = a_0 + a_1(w - z_0) + \cdots + a_n(w - z_0)^n + \cdots , \qquad (10)$$

where the coefficients a_n are given by

$$a_n = \frac{1}{2\pi i} \int_{K_1} \frac{f(\zeta)\,d\zeta}{(\zeta - z_0)^{n+1}}. \qquad (11)$$

We now define

$$f_+(w) = h_1(w).$$

We now consider the integral $h_2(w)$. If the point z_0 belongs to the open set G, the function $h_2(w)$ is identically zero, by Cauchy's integral theorem, since the function $f(z)$ and hence the function $g(z)$ are differentiable throughout an open set that contains the circle K_2 and the entire interior of K_2. In this case, we have the identity

$$f(w) = h_1(w) = f_+(w):$$

see the equalities (7) and (8).

Suppose now that z_0 does not belong to the open set G. The integrand in the integral defining $h_2(w)$ can be written in the form

$$\frac{f(\zeta)}{\zeta - w} = -\frac{f(\zeta)}{(w - z_0) - (\zeta - z_0)} = -\frac{f(\zeta)}{w - z_0} \cdot \frac{1}{1 - \dfrac{\zeta - z_0}{w - z_0}}$$

$$= -\sum_{k=0}^{\infty} f(\zeta) \frac{(\zeta - z_0)^k}{(w - z_0)^{k+1}}. \qquad (12)$$

Here we have expanded the expression

$$\frac{1}{1 - \dfrac{\zeta - z_0}{w - z_0}}$$

as a geometric progression, which can be done inasmuch as

$$\left| \frac{\zeta - z_0}{w - z_0} \right| = \sigma < 1$$

for all ζ on the circle K_2. (Observe that ζ lies on the circle K_2 while w is in the exterior of K_2.) Therefore the series in the second line of (12) converges uniformly and so can be integrated term by term. We obtain

$$-h_2(w) = \frac{a_{-1}}{w - z_0} + \frac{a_{-2}}{(w - z_0)^2} + \cdots + \frac{a_{-k}}{(w - z_0)^k} + \cdots , \qquad (13)$$

where

$$a_{-k} = \frac{1}{2\pi i} \int_{K_2} f(\zeta)(\zeta - z_0)^{k-1} d\zeta. \tag{14}$$

We now set

$$f_-(w) = -h_2(w).$$

It is possible that all of the coefficients $a_{-1}, a_{-2}, \ldots$ vanish even though z_0 does not belong to the set G. If this occurs, of course, the function $f_-(w)$ is identically zero, and we have

$$f(w) = f_+(w).$$

That is, we may define $f(z_0)$ as the value a_0 and obtain a function that is differentiable in the open set G to which we have adjoined the point z_0.

We have thus achieved the expansion of the function $f(w)$ as the sum of the two function $f_+(w)$ and $f_-(w)$, each expansible as a power series as in (1) and (3) respectively. We have written the variable as w instead of z, but this is a trivial matter: we simply replace "w" by "z".

It is worth while observing that the series (13) (or (3)) for the function $f_-(w)$ converges for *all* values of $w \neq z_0$. Our construction shows that the series (13) converges for all w such that

$$r_2 < |w - z_0| < r'.$$

It was already observed in §32 that the coefficients a_{-k} do not depend on the choice of the radius r_2 of the circle K_2. Therefore (13) converges for all w such that

$$0 < |w - z_0| < r_1.$$

The series (13) converges also for all w lying outside of the closed disk bounded by K_1, because for all such w, the terms of the series (13) are smaller in modulus term by term. Thus the series (3) and (13) converge under the restriction (4).

If the point z_0 does not belong to G, we call it *an isolated singular point of the function* $f(z)$. Isolated singular points may be classified as follows. If the series (3) has an infinite number of nonzero coefficients a_{-k}, z_0 is said to be *an essential singularity of the function* $f(z)$. If at least one of the coefficients a_{-k} is different from zero, and only a finite number of them are different from zero, the point z_0 is called *a pole of the function* $f(z)$. The largest positive integer k such that $a_{-k} \neq 0$ is called *the order of the pole*. If all of the coefficients in the series (3) are zero, the point z_0 is called *a removable singularity of the function* $f(z)$. If we define $f(z_0)$ as a_0, we find that the function $f(z)$ is in fact differentiable in the open set obtained by adjoining the point z_0 to the open set G.

We have already used the term "analytic" in connection with functions $f(z)$. A function $f(z)$ can be said to be *analytic in an open set* G if it is defined throughout G and at every point z_0 of G admits a Taylor series expansion of the form (1) with positive radius of convergence. A special case of what we have just proved is that a differentiable function $f(z)$ defined in an open set G is analytic in G.

Let us summarize our results about expansion of a function $f(z)$ in a Laurent series in a neighborhood of a point z_0. The function admits an expansion of the form

$$f(z) = \sum_{n=-\infty}^{\infty} a_n(z-z_0)^n = \lim_{k \to \infty} \sum_{n=-k}^{k} a_n(z-z_0)^n. \tag{15}$$

It is natural to ask if the coefficients a_n in the series expansion (15) are determined uniquely by the function $f(z)$. In other terms, is it conceivable that a function $f(z)$ could admit two different Laurent series expansions about a point z_0? We will show that this cannot in fact occur, in that we can compute the coefficients a_n once we know the function $f(z)$. We observe first that the series (15) converges uniformly on every circle K with center z_0 and radius not exceeding r_1. Let us multiply the series (15) by the function $(z-z_0)^{-(k+1)}$, where k is an arbitrary integer, positive, negative, or zero. We get the series

$$f(z) \cdot (z-z_0)^{-(k+1)} = \sum_{n=-\infty}^{\infty} a_n(z-z_0)^{n-(k+1)}. \tag{16}$$

Like the series in (15), the series in (16) converges uniformly on the circle K. Thus we may integrate both sides of the identity (16) and on the right we may integrate term by term. We find that

$$\int_K f(\zeta)(\zeta-z_0)^{-(k+1)}d\zeta = \sum_{n=-\infty}^{\infty} a_n \int_K (\zeta-z_0)^{n-(k+1)}d\zeta. \tag{17}$$

We have computed the integrals appearing on the right side of (17), in §32, formulas (24) and (25), obtaining the result that $\int_K (\zeta-z_0)^{n-(k+1)}d\zeta$ vanishes except for the case $n-(k+1)=-1$, that is, for $n=k$, and for $n=k$ it is equal to $2\pi i$. From the identity (17) we therefore find that

$$a_k = \frac{1}{2\pi i} \int_K f(\zeta)(\zeta-z_0)^{-(k+1)}d\zeta. \tag{18}$$

The formula (18) gives a_k explicitly in terms of the function $f(z)$. That is, the coefficients appearing in the Laurent expansion (15) are completely determined by the function $f(z)$.

We turn next to another uniqueness problem.

Example 1. A connected open set G in the complex plane will be called a *region*. We recall from §23 that an open set G is connected if every pair of points in G can be connected by a broken line lying entirely within G. Let $f_1(z)$ and $f_2(z)$ be two analytic functions defined in a region G. Suppose that there is an infinite subset M of G admitting some limit point z_0 in the set G such that $f_1(z) = f_2(z)$ for all z in the set M. We will prove that $f_1(z)$ and $f_2(z)$ are necessarily identically equal throughout G.

Let z^* be an arbitrary point of the region G and let $z = z(t)$ be a parametric equation for a path lying wholly in G that connects z_0 with z^*. That is, we have

$$z(t_0) = z_0 \quad \text{and} \quad z(t^*) = z^*, \qquad t_0 \leq t \leq t^*. \tag{19}$$

We proved in §20 that if $f_1(z) = f_2(z)$ for all z in M, then $f_1(z)$ and $f_2(z)$ have the same Taylor series expansions at the point z_0. Let r be the (positive) radius of convergence of this Taylor series. Now consider the path (19), which begins at the point $z_0 = z(t_0)$. For all t close enough to t_0, this path lies within the disk of radius r and center at z_0. For all such t, we accordingly have

$$f_1(z(t)) = f_2(z(t)). \tag{20}$$

Now let t_1 be the supremum of the set of all t's in the closed interval $[t_0\, t^*]$ for which (20) holds for all t between t_0 and t_1. Since the functions $f_1(z(t))$ and $f_2(z(t))$ are continuous functions of t, we see at once that

$$f_1(z(t_1)) = f_2(z(t_1)). \tag{21}$$

Furthermore, the point $z(t_1)$ is a limit point of the infinite set of points $z(t)$, $t_0 \leq t < t_1$, where $f_1(z(t)) = f_2(z(t))$, and so the argument just gone through shows that $f_1(z)$ and $f_2(z)$ have the same Taylor series expansion at the point $z_1 = z(t_1)$. Let r_1 be the (positive) radius of convergence of this Taylor series. If we assume that $t_1 < t^*$, then, since $z(t)$ is continuous, there is a parameter interval $t_1 \leq t \leq t_2$ with $t_2 > t_1$ such that $f_1(z(t)) = f_2(z(t))$ for all t such that $t_1 \leq t \leq t_2$. This contradicts the definition of t_1 as the supremum of the set of t's such that $f_1(z(t)) = f_2(z(t))$ for all t such that $t_0 \leq t \leq t_1$. It follows that $f_1(z^*) = f_2(z^*)$, which is to say that $f_1(z)$ and $f_2(z)$ are in fact the same function throughout G.

A special case of this result is the following. If the analytic function $f(z)$ in G vanishes everywhere on the set M, it is identically zero throughout G.

Example 2. Our definition of the derivative of a function of a real variable is formally the same as our definition of the derivative of a function of a complex variable. See §21 for the details. This made it possible for us to develop the two concepts simultaneously. Nevertheless, the real significance of the two concepts is very different indeed. The requirement that a function of a real variable admit a derivative at every point is fairly weak. The requirement that a function of a complex variable be differentiable at every point, on the other hand, is an extraordinarily strong demand. Thus, we have an example of a function of a real variable that is defined for all real numbers, has derivatives of all orders at every point, and yet at the point 0 cannot be expanded in a power series: see Example 2 in §26. By contrast, a function of a complex variable that has only one derivative at every point can be expanded in a power series. The fundamental difference between the two cases is the following. Suppose that we have a function $f(x)$ of a real variable, and that we are interested in whether or not it has a derivative at a point x. We study the quotient

$$\frac{f(\xi) - f(x)}{\xi - x}$$

as $\xi < x$. The point ξ can approach x only from the left or from the right. If we have a function $f(z)$ of a complex variable, on the other hand, and wish to study its differentiability at a point z, we must look at the quotients

$$\frac{f(\zeta) - f(z)}{\zeta - z} \tag{22}$$

for all complex ζ in a disk of small positive radius with center at z. Thus ζ can converge to z from any direction, e.g. horizontally or vertically or along some complicated path maintaining no constant direction at all. We must achieve the same limit for all such modes of approach, and this lays an exceedingly strong restriction on the function $f(\zeta)$. We consider two particular modes of approach. Let us write $z = x + iy$. We write $f(z) = u(x, y) + iv(x, y)$, where $u(x, y)$ and $v(x, y)$ are real-valued functions of the complex variable z, or, what is the same thing, the two real variables x and y. We also write $\zeta = \xi + i\eta$. Then the difference quotient (22) is equal to

$$\frac{u(\xi, \eta) + iv(\xi, \eta) - u(x, y) - iv(x, y)}{(\xi - x) + i(\eta - y)}. \tag{23}$$

Suppose now that $\zeta \to z$ horizontally. In (23), we must set $\eta - y = 0$, and in the limit as $\zeta \to z$, (23) becomes

$$\frac{\partial u(x, y)}{\partial x} + i\frac{\partial v(x, y)}{\partial x}. \tag{24}$$

Suppose next that $\zeta \to z$ vertically. Then in (23) we must set $\xi - x = 0$, and the limit of (23) is

$$\frac{1}{i}\frac{\partial u(x, y)}{\partial y} + \frac{\partial v(x, y)}{\partial y}. \tag{25}$$

The expressions (24) and (25) are equal if $f(z)$ is differentiable at z, and so we find that

$$\frac{\partial u(x, y)}{\partial x} = \frac{\partial v(x, y)}{\partial y} \quad \text{and} \quad \frac{\partial u(x, y)}{\partial y} = \frac{\partial v(x, y)}{\partial x}.$$

These equations are called *the Cauchy-Riemann equations* for the real and imaginary parts of an analytic function of a complex variable. They illustrate the extreme demand that differentiability for a function of a complex variable actually imposes.

§34. Residues

We consider an analytic function $f(z)$ defined throughout a region G in the complex plane. As we made clear in the preceding section, it may well happen that there are open disks that, except for their centers, lie entirely within G. We call the center z_0 of such a disk an isolated singular point of $f(z)$, just as in the preceding section. We divided isolated singular points into three classes: removable singularities, essential singularities, and poles. We can define $f(z_0)$ for a removable singularity so that the resulting function is analytic in G with z_0 added to G, and so it is natural to suppose that G actually contains all removable singularities. That is, we remove all removable singularities. It is also a convenience in the other two cases to regard these singular points as being in G, although of course $f(z)$ is not defined at nonremovable singular points. We write the set G with all of the isolated singular points of $f(z)$ added to it again by the symbol G. We will say that the analytic function $f(z)$ is given in the region G (after we add the isolated singularities to the connected open set G, we still have a connected open set). The set M of isolated singularities of the function $f(z)$ may well be infinite, but is cannot have a limit point in the set G. For, given any point z_0 of G, the function $f(z)$ is differentiable in a certain open disk with center at z_0 (possibly except for the center z_0). Within this "punctured" open disk the function $f(z)$ has no singularities. Consequently z_0 cannot be a limit point of the set M.

Behavior of an Analytic Function in the Neighborhood of a Pole or Essential Singularity. Let $f(z)$ be an analytic function given in a region G and let z_0 be a point of G that is not an essential singularity of the function $f(z)$. (The point z_0 may be a pole or it may be a point of analyticity.) There exists an integer μ such that in some open disk of positive radius with center at z_0 the function $f(z)$ can be written in the form

$$f(z) = (z - z_0)^\mu \hat{f}(z), \tag{1}$$

where $\hat{f}(z)$ admits a Taylor series expansion with positive radius of convergence,

$$\hat{f}(z) = \hat{a}_0 + \hat{a}_1(z - z_0) + \cdots + \hat{a}_n(z - z_0)^n + \cdots, \tag{2}$$

with the property that

$$\hat{a}_0 = \hat{f}(z_0) \neq 0.$$

(The sole exception to this assertion is the case in which $f(z)$ is the function identically zero in G.)

Plainly the function identically zero does not admit an expansion of the forms (1) and (2). Suppose then that $f(z)$ is not the function identically zero. Since z_0 is not an essential singularity of $f(z)$, the Laurent series for $f(z)$ in some set consisting of all z's of the form $0 < |z - z_0| < r \ (r > 0)$ has the form

$$f(z) = \sum_{n=\mu}^{\infty} a_n (z - z_0)^n. \tag{3}$$

We determine μ by the condition that a_μ be different from 0. This a_μ exists for $f(z)$ not identically zero in view of the results established in the preceding section.

The expansion (3) is equivalent to

$$f(z) = (z - z_0)^\mu [a_\mu + a_{\mu+1}(z - z_0) + \cdots + a_{\mu+n}(z - z_0)^n + \cdots]. \tag{4}$$

This proves (2).

If μ is negative, the function $f(z)$ has a pole of order $-\mu$ at z_0, as defined in the preceding section. If μ is positive, we say that $f(z)$ *has a zero of order μ at z_0*.

The Quotient of two Analytic Functions. Let $\hat{\varphi}(z)$ and $\hat{\psi}(z)$ be two analytic functions given throughout a region G, let z_0 be a non-singular point for both of them, and let

$$\hat{\varphi}(z) = \hat{\alpha}_0 + \hat{\alpha}_1(z - z_0) + \cdots + \hat{\alpha}_n(z - z_0)^n + \cdots, \tag{5}$$

$$\hat{\psi}(z) = \hat{\beta}_0 + \hat{\beta}_1(z - z_0) + \cdots + \hat{\beta}_n(z - z_0)^n + \cdots \tag{6}$$

be the expansions of these two functions in Taylor series about the point z_0. Suppose also that

$$\hat{\alpha}_0 = \hat{\varphi}(z_0) \neq 0 \quad \text{and} \quad \hat{\beta}_0 = \hat{\psi}(z_0) \neq 0. \tag{7}$$

Let ρ be the radius of convergence of the series (5), σ the radius of convergence of the series (6), and σ' the distance from z_0 to the nearest zero of the function $\hat{\psi}(z)$. Let σ'' be the smallest of the numbers ρ, σ, and σ'. Then the quotient function $\hat{\chi}(z) = \dfrac{\hat{\varphi}(z)}{\hat{\psi}(z)}$ admits a Taylor series expansion

$$\frac{\hat{\alpha}_0}{\hat{\beta}_0} + \hat{\gamma}_1(z - z_0) + \cdots + \hat{\gamma}_n(z - z_0)^n + \cdots. \tag{8}$$

The radius of convergence r of the series (8) is greater than or equal to σ''.

We will prove this assertion. As we proved in §22, (17), the quotient $\hat{\varphi}(z)/\hat{\psi}(z)$ is differentiable at every point z such that $|z - z_0| < \sigma''$. Therefore it is an analytic function in this region, and by the results of the preceding section it admits a Taylor series expansion about the point z_0 whose radius of convergence is no smaller than σ''. The first term of the Taylor series expansion (8) is equal to

$$\hat\chi(z_0)=\hat\phi(z_0)/\hat\psi(z_0)=\hat\alpha_0/\hat\beta_0\neq0. \tag{9}$$

This completes the proof.

Now let $\varphi(z)$ and $\psi(z)$ be two functions which are defined in a neighborhood of z_0 by series of the form (4):

$$\varphi(z)=(z-z_0)^p\,\hat\phi(z)\quad\text{and}\quad\psi(z)=(z-z_0)^q\,\hat\psi(z),$$

where the functions $\hat\phi(z)$ and $\hat\psi(z)$ are expansible in Taylor series in a neighborhood of z_0, and where

$$\hat\phi(z_0)\neq0\quad\text{and}\quad\hat\psi(z_0)\neq0.$$

We can then write the quotient $\chi(z)=\varphi(z)/\psi(z)$ in the form

$$\chi(z)=\frac{\varphi(z)}{\psi(z)}=(z-z_0)^{p-q}\,\hat\chi(z), \tag{10}$$

where

$$\hat\chi(z_0)=\frac{\hat\phi(z_0)}{\hat\psi(z_0)}\neq0. \tag{11}$$

This assertion follows at once from (8) and (9).

Residues. Let $f(z)$ be an analytic function defined in a region G and let z_0 be a point of the region. We define

$$v(z_0)=\frac{1}{2\pi i}\int_K f(\zeta)\,d\zeta, \tag{12}$$

where K is a circle with center at z_0 with radius greater than 0 but so small that the value of the integral is independent of the radius. (In particular, K and its interior must lie in a region in which z_0 is the only possible singularity of $f(z)$: and z_0 itself need not be a singularity of $f(z)$.) If z_0 is not a singularity of $f(z)$, then Cauchy's integral theorem (§32, (20) shows that $v(z_0)=0$. If z_0 is a singularity of $f(z)$, the number $v(z_0)$ is given by the equality

$$v(z_0)=a_{-1},$$

where a_{-1} is the coefficient of $(z-z_0)^{-1}$ in the Laurent series expansion of $f(z)$ at the point z_0 (see §33, (14)). The number $v(z_0)$ is called *the residue of $f(z)$ at the point z_0*.

Now let K_0 be a closed curve without self-intersections such that all of the points on K_0 or interior to K_0 belong to the region G. Then within K_0 and on K_0 there can be only a finite number of singularities of the function $f(z)$. For, if there were an infinite number of singularities, they would all lie in the closed bounded set consisting of K_0 together with its interior, and would admit a limit point that would be in G, contrary to our hypotheses about G and $f(z)$. We also suppose that there are no singularities of $f(z)$ actually on the curve K_0. Let $z_1, z_2, \ldots, z_m$ be all of the singularities of $f(z)$ lying in the interior of K_0. Then we have

$$\frac{1}{2\pi i} \int_{K_0} f(\zeta)\,d\zeta = v(z_1) + v(z_2) + \cdots + v(z_m). \tag{13}$$

To prove (13), we construct circles K_j about all of the points z_j having very small radii. If these radii are sufficiently small, the hypotheses for formula (30) of §32 will be satisfied, and we get

$$\frac{1}{2\pi i} \int_{K_0} f(\zeta)\,d\zeta = \sum_{j=1}^{m} \frac{1}{2\pi i} \int_{K_j} f(\zeta)\,d\zeta = \sum_{j=1}^{m} f(z_j).$$

This is exactly (13).

Logarithmic Residues. Let $f(z)$ again be an analytic function defined in G. The function

$$g(z) = \frac{f'(z)}{f(z)}$$

is called *the logarithmic derivative of the function $f(z)$*, since

$$g(z) = \frac{d}{dz} \ln f(z).$$

(Actually we do not need the function $\ln f(z)$: we mention it only to explain our terminology.)

Let z_0 be any point of the region G that is not an essential singularity of the function $f(z)$. Let us write the function $f(z)$ in a neighborhood of z_0 in the form (4). We may differentiate the series (4) term by term, obtaining

$$f'(z) = \mu \hat{a}_0 (z - z_0)^{\mu - 1} + (\mu + 1)\hat{a}_1 (z - z_0)^{\mu} + \cdots$$
$$+ (\mu + n)\hat{a}_n (z - z_0)^{\mu + n - 1} + \cdots .$$

As in (10), it follows that

$$g(z) = \frac{f'(z)}{f(z)} = \mu(z - z_0)^{-1} + \gamma(z), \tag{14}$$

where $\gamma(z)$ is expansible in a Taylor series about z_0 (with nonnegative powers of $z-z_0$). Therefore the residue of the logarithmic derivative of $f(z)$ (*the logarithmic residue of* $f(z)$) is defined by the formula

$$\frac{1}{2\pi i}\int_K f(\zeta)\,d\zeta=\mu.$$

We now consider a function $f(z)$ defined, analytic, and not identically zero in a region G, and a closed curve K_0 without self-intersections that lies in G, together with all of its interior. We suppose that there are no essential singularities of the function $f(z)$ either on K_0 or in its interior. Finally we suppose that there are neither zeros nor poles of $f(z)$ on the curve K_0 itself. Let $z_1, z_2, ..., z_p$ be the set of all zeros of $f(z)$ that lie in the interior of K_0, and suppose that z_j has order μ_j. Let $z_1^*, z_2^*, ..., z_q^*$ be the set of all poles of $f(z)$ lying in the interior of K_0, and suppose that z_k^* has order ν_k. (Recall that the numbers μ_j and ν_k are positive integers.) Finally let $h(z)$ be a function that is defined and analytic throughout G and admits no singularities at all, either poles or essential singularities, throughout G. We then have the following very important identity:

$$\frac{1}{2\pi i}\int_{K_0} h(\zeta)\frac{f'(\zeta)}{f(\zeta)}\,d\zeta=\sum_{j=1}^{p}\mu_j h(z_j)-\sum_{k=1}^{q}\nu_k h(z_k^*).\tag{15}$$

To prove (15), we let K_j be a circle with small radius centered at z_j and L_k a circle with small radius centered at z_k^*. From (14) and from (28) in §32, we obtain

$$\frac{1}{2\pi i}\int_{K_j} h(\zeta)\frac{f'(\zeta)}{f(\zeta)}\,d\zeta=\frac{1}{2\pi i}\int_{K_j} h(\zeta)\frac{\mu_j d\zeta}{\zeta-z_j}=\mu_j h(z_j).$$

The same considerations give us

$$\frac{1}{2\pi i}\int_{L_k} h(\zeta)\frac{f'(\zeta)}{f(\zeta)}\,d\zeta=\frac{1}{2\pi i}\int_{L_k} h(\zeta)\frac{(-\nu_k)}{\zeta-z_k^*}\,d\zeta=-\nu_k h(z_k^*).$$

Applying formula (30) of §32 to the last two identities, we obtain (15).

Suppose in particular that $h(z)$ is the function identically 1. We find that

$$\frac{1}{2\pi i}\int_{K_0}\frac{f'(\zeta)}{f(\zeta)}\,d\zeta=\mu-\nu,\tag{16}$$

where μ is the number of zeros of $f(z)$ within K_0, each counted with its order, and ν is the number of poles of $f(z)$ within K_0, each counted with its order.

Another important case is that in which $h(z)=z$. We obtain

$$\frac{1}{2\pi i}\int_{K_0}\frac{\zeta f'(\zeta)}{f(\zeta)}\,d\zeta=\sum_{j=1}^{p}\mu_j z_j-\sum_{k=1}^{q}\nu_k z_k^*.\tag{17}$$

Example. We can prove the fundamental theorem of algebra by using the logarithmic residue. That is, we can prove that a complex polynomial $f(z) = z^n + a_1 z^{n-1} + \cdots + a_n$ admits exactly n complex roots, each counted with its order. We will do this by forming the logarithmic derivative

$$g(z) = \frac{f'(z)}{f(z)} = \frac{n z^{n-1} + (n-1) a_1 z^{n-2} + \cdots + a_{n-1}}{z^n + a_1 z^{n-1} + \cdots + a_n} \tag{18}$$

and showing that

$$\frac{1}{2\pi i} \int_K g(\zeta) \, d\zeta = n \tag{19}$$

for all circles K with center at $z=0$ and sufficiently large radius r.

The integral on the left side of the desired identity (19) is taken over a curve where the modulus is large. It is therefore reasonable to make a change of variable in our definition (18) of $g(z)$, namely,

$$z = \frac{1}{u}. \tag{20}$$

We make the change of variable (20) in both ends of (18) and then multiply both numerator and denominator by u^n. We obtain

$$g\left(\frac{1}{u}\right) = \frac{n u + u \, \varphi(u)}{1 + \psi(u)}, \tag{21}$$

where the functions $\varphi(u)$ and $\psi(u)$ are polynomials in the variable u, both with constant term zero. The function $g\left(\frac{1}{u}\right)$ accordingly admits a power series expansion in terms of u:

$$g\left(\frac{1}{u}\right) = n u + \gamma_1 u^2 + \cdots + \gamma_n u^{n+1} + \cdots \tag{22}$$

having positive radius of convergence ρ.

We now write $1/z$ instead of u in (22), finding that

$$g(z) = \frac{n}{z} + \frac{\gamma_1}{z^2} + \cdots + \frac{\gamma_n}{z^{n+1}} + \cdots. \tag{23}$$

For $|z| = r > \dfrac{1}{\rho}$, the series (23) converges uniformly. It can therefore be integrated term by term over the circle K of radius r. This gives us (19), as we wished to prove.

A polynomial $f(z)$ has no poles at all, and so (16) and (19) show that the polynomial $f(z)$ has exactly n zeros (each counted with its order) within the circle K of radius r.

§35. Finding Inverse Functions

We will use the results of §34 to find inverse functions, defined as in §22. Consider an equation

$$\psi(w) = z, \qquad (1)$$

where $\psi(w)$ is an analytic function of the complex variable w, admitting a Taylor series expansion in a neighborhood of the point w_0. That is, $\psi(w)$ can be written as a power series in nonnegative powers of $w - z_0$. We write $\psi(w_0) = z_0$ and we suppose that $\psi'(w_0) \neq 0$. In the equation (1), we take z to be a known quantity and w to be the unknown for which we wish to solve. We will prove that the equation (1) admits a solution $w = \varphi(z)$, where $\varphi(z_0) = w_0$ and $\varphi(z)$ admits a Taylor series expansion in (nonnegative) powers of the variable $z - z_0$ having positive radius of convergence.

To prove this, we first consider a circle K in the W-plane with center w_0 and with radius so small that the only solution of the equation $\psi(w) - \psi(w_0) = 0$ with w on the circle K or in the interior of the circle K is $w = w_0$. This zero has order 1, since $\psi'(w_0) \neq 0$ (see §20). Therefore there is a positive number δ such that

$$|\psi(w) - \psi(w_0)| > \delta > 0 \qquad (2)$$

for all points w on the circle K. Formula (16) of §34 shows that

$$\frac{1}{2\pi i} \int_K \frac{\psi'(\omega)}{\psi(w) - \psi(w_0)} \, d\omega = 1. \qquad (3)$$

Formula (17) of §34 shows that

$$\frac{1}{2\pi i} \int_K \frac{\omega \psi'(\omega)}{\psi(\omega) - \psi(w_0)} \, d\omega = w_0. \qquad (4)$$

We will now look for a zero $\varphi(z)$ of the function

$$\psi(w) - z = (\psi(w) - \psi(w_0)) - (z - z_0) \qquad (5)$$

in the interior of the circle K for sufficiently small values of $|z - z_0|$. In view of formula (16) of §34, the number

$$\frac{1}{2\pi i} \int_K \frac{\psi'(\omega)}{\psi(\omega) - z} \, d\omega$$

is an integer. For $z = z_0$, (3) shows that this integer is 1. Considerations of continuity show that this integer remains 1 for all z such that $|z - z_0|$ is sufficiently small. Therefore the the equation (1) admits exactly one solution in the interior of the circle K if $|z - z_0|$ is sufficiently small. We can compute the zero $\varphi(z)$ of the function $\psi(w) - z$ from formula (17) of §34 (see also (5)):

$$\varphi(z) = \frac{1}{2\pi i} \int_K \frac{\omega \psi'(\omega)}{(\psi(\omega) - \psi(w_0)) - (z - z_0)}\, d\omega. \tag{6}$$

We rewrite the integrand in (6):

$$\frac{\omega \psi'(\omega)}{(\psi(\omega) - \psi(w_0)) - (z - z_0)} = \frac{\omega \psi'(\omega)}{(\psi(\omega) - \psi(w_0))\left[1 - \dfrac{z - z_0}{\psi(\omega) - \psi(w_0)}\right]}$$

$$= \sum_{n=0}^{\infty} \omega \psi'(\omega) \frac{(z - z_0)^n}{[\psi(\omega) - \psi(w_0)]^{n+1}}. \tag{7}$$

We obtain the last line of (7) by expanding the function

$$\frac{1}{1 - \dfrac{z - z_0}{\psi(\omega) - \psi(w_0)}}$$

in a geometric progression, which converges for z such that $|z - z_0| < \delta$ and for ω lying on the circle K (see the inequalities (2)). The series in (7) converges uniformly on the circle K and so can be integrated term by term. We find accordingly from (6) and (7) that

$$\varphi(z) = \sum_{n=0}^{\infty} (z - z_0)^n \cdot \frac{1}{2\pi i} \int_K \frac{\omega \psi'(\omega)}{[\psi(\omega) - \psi(w_0)]^{n+1}}\, d\omega. \tag{8}$$

The power series on the right side of (8) converges for all z such that $|z - z_0| < \delta$. From (4) we see that the constant term in the power series in (8) is equal to w_0. Thus we have found a solution of the equation (1) in the form of a power series in the variable $z - z_0$. Now let us suppose that the function $\psi(w)$, defined in a neighborhood of w_0 by its Taylor series, is actually defined in some region H of the W-plane. Let us also suppose that the function $\varphi(z)$, defined by its power series expansion (8) in some neighborhood of z_0, is actually defined in some region G of the Z-plane. Suppose finally that φ maps every point of the region G onto a point in the region H. Now form the function $\psi(\varphi(z))$. This function is analytic for all z that belong to the region G, since we can compute its derivative for all such z by the composite function rule. For all values of z that are near enough to z_0, we have the identity

$$\psi(\varphi(z)) = z. \tag{9}$$

The functions on the left and right sides of (9) are analytic in G and are equal for all values of z that are sufficiently close to z_0. It follows from §33 that these functions agree throughout G, which is to say that

$$\psi(\varphi(z)) \equiv z \tag{10}$$

for all z in G.

Example 1. Consider the equation

$$e^w = 1 + z. \tag{11}$$

A solution of this equation is the multiple-valued function $\ln(1+z)$: we write $\varphi(z) = \ln(1+z)$. Suppose that $\varphi(z)$ exists and is analytic. Differentiating the identity

$$e^{\varphi(z)} = 1 + z, \tag{12}$$

we find that $\varphi'(z) e^{\varphi(z)} = 1$, so that (12) yields

$$\varphi'(z) = \frac{1}{1+z}.$$

Expand the right side of this identity in a geometric progression to obtain

$$\varphi'(z) = 1 - z + z^2 - \cdots + (-1)^n z^n + \cdots. \tag{13}$$

Integrate (13) term by term and suppose further that $\varphi(0) = 0$. We find that

$$\varphi(z) = z - \frac{z^2}{2} + \frac{z^3}{3} - \cdots + (-1)^n \cdot \frac{z^{n+1}}{n+1} + \cdots. \tag{14}$$

The series on the right side of (14) has radius of convergence 1. It appears at first glance that (14) gives us a power series expansion of the function $\ln(1+z)$. However this is not so. We proceeded from the assumption that $\varphi(z)$ is a differentiable function satisfying (12). But we do not know *a priori* that $\varphi(z)$, differentiable and satisfying (12), exists. All we have proved is that if $\varphi(z)$ exists, then it has the power series expansion (14). We would have to show that if we put the power series (14) into the equation (12), then a true identity results. This is by no means easy to accomplish.

Another course of action, however, is simple enough. Let us suppose that the power series (14) does yield the function $\ln(1+z)$. Then the derivative $\ln'(1+z)$ has the power series expansion (13) and so $\ln'(1+z) = \frac{1}{1+z}$. That is, we can prove that if (14) gives the function $\ln(1+z)$, then we must have $\ln'(u) = \frac{1}{u}$. From this it is easy to show that $\ln e^w = w$. In fact, compute the derivative of the function $\ln u$, with the proviso that $u=e^w$. We obtain

$$\frac{d}{dw}\ln u = \ln' u \frac{du}{dw} = \frac{1}{u}\frac{d}{dw}e^w = \frac{1}{u}u = 1.$$

This yields

$$\frac{d}{dw}\ln e^w = 1,$$

from which it follows that

$$\ln e^w = w + c.$$

In this identity set $w=0$ and suppose that $\ln 1 = 0$. We find that $c=0$, so that

$$\ln e^w = w.$$

We proved this identity from the assumption that $\ln(1+z)$ is given by the power series (14). This is by no means a proof, however, that the function $\varphi(z)$ defined by (14) is a solution of the equation (12). We do not even know that a solution to the equation (12) exists. This gap is filled by the result of the present section.

To show that the equation (11) admits a solution, we use our results on the solvability of the equation (1). We write $\psi(w)=e^w-1$ and for w_0 we take the point $w=0$. Then (11) has the form

$$\psi(w)=z \tag{15}$$

with $\psi(0)=0$ and $\psi'(0)=1$. We have the solution $\varphi(z)$ for small values of z in the form (8):

$$\varphi(z)=a_1 z + a_2 z^2 + \cdots + a_n z^n + \cdots. \tag{16}$$

The power series in (16) satisfies the equation $e^{\varphi(z)}=1+z$ for small values of z and so, for small values of z, must have the form (14). That is, the coefficients of the power series in (16) are the same as the coefficients of the power series in (14). Thus the function $\varphi(z)$ defined by (16) is defined for all z such that $|z|<1$. The function $\psi(w)$ is defined and differentiable for all complex w. Accordingly (10) shows that

$$\psi(\varphi(z))\equiv z$$

for all z such that $|z|<1$. That is, the series (14) yields a value of $\ln(1+z)$ for all z such that $|z|<1$.

Example 2. The situation is completely analogous for the function $\arcsin z$, which we define as a solution $w=\varphi(z)$ of the equation $\sin w = z$. Let us suppose that there is a differentiable function $\varphi(z)$ satisfying the equation

$$\sin \varphi(z) = z. \tag{17}$$

Then we show without any difficulty that

$$\varphi'(z) = (1 - z^2)^{-\frac{1}{2}},$$

as in (25) of §22. From this we obtain the power series expansion given in (26), §26 for the function $\varphi(z)$. However, to prove that a function $\varphi(z)$ satisfying (17) exists, we must make use of the result of the present section. Just as we did for the logarithm, we can show that the function $\varphi(z)$ defined by the power series (26) of §26 satisfies the identity

$$\sin \varphi(z) = z \tag{18}$$

for all z such that $|z| < 1$.

It is also very easy to prove that the identity $\varphi(\sin w) = w$ holds for our function $\varphi(z)$: proceed as we did with the function $\ln u$. At the same time, there is evidently no simple proof that the two identities are equivalent.

§36. Entire Functions and Singular Points

A function $f(z)$ of a complex variable that is defined and differentiable throughout the whole complex plane is said to be *entire*. The results of §33 show that an entire function can be expanded in a Taylor series about any point of the complex plane, and that this Taylor series converges everywhere. In particular, we may expand about the point 0, finding that an entire function $f(z)$ can be written in the form

$$f(z) = a_0 + a_1 z + \cdots + a_n z^n + \cdots, \tag{1}$$

where the series in (1) converges (absolutely) for all z. If the series (1) admits only a finite number of nonzero coefficients a_k, the function $f(z)$ is a polynomial. However, we have more complicated examples of entire functions already at hand: these are the functions $e^z, \sin z$, and $\cos z$, which we studied in §19.

Bounded Entire Functions. Let $f(z)$ be a bounded entire function, that is, for which there is a positive number c such that $|f(z)| < c$ for all complex numbers z. We will prove that $f(z)$ is a constant:

$$f(z) = c_1 \tag{2}$$

for some constant c_1 and all complex z.

We use formula (18) of §33 to prove this. For a circle K of positive radius r and center at 0, we have

$$a_n = \frac{1}{2\pi i} \int_K f(\zeta)\, \zeta^{-(n+1)}\, d\zeta.$$

On the circle K, we have

$$|f(\zeta)\, \zeta^{-(n+1)}| < \frac{c}{r^{n+1}}.$$

The estimate (22) of §31, p. 266, shows that

$$|a_n| = \frac{1}{2\pi} \left| \int_K f(\zeta)\, \zeta^{-(n+1)}\, d\zeta \right| \leq \frac{1}{2\pi} \frac{c}{r^{n+1}} \cdot 2\pi r = \frac{c}{r^n}. \tag{3}$$

The coefficient a_n does not depend upon r. For $n \geq 1$, the right end of (3) goes to zero as $r \to \infty$. It follows that $f(z) = a_0$ for all z: and this is what we claim in (2).

The Behavior of a Polynomial as $z \to \infty$. Consider the special case in which the series (1) has only a finite number of nonzero terms:

$$f(z) = a_0 + a_1 z + \cdots + a_n z^n;$$

we agree that $a_n \neq 0$ and consider only the case $n \geq 1$. We can then write

$$f(z) = z^n \left(a_n + \frac{a_{n-1}}{z} + \cdots + \frac{a_0}{z^n} \right).$$

From this it follows that $|f(z)| \to \infty$ as $|z| \to \infty$. We write

$$\lim_{z \to \infty} f(z) = \infty.$$

Nonpolynomial Entire Functions. If the entire function $f(z)$ fails to be a polynomial, its power series expansion (1) contains an infinite number of nonzero coefficients. Such a function enjoys the following remarkable property. Let a be an absolutely arbitrary complex number. There exists a sequence of complex numbers

$$z_1, z_2, \ldots, z_n, \ldots$$

for which

$$\lim_{n \to \infty} |z_n| = \infty \tag{4}$$

and for which

$$\lim_{n \to \infty} f(z_n) = a. \tag{5}$$

We prove this fact by contradiction. Its denial may be stated as follows. There is a complex number a and there is a positive number ε such that for some positive number ρ, the inequality

$$|f(z)-a|>\varepsilon \text{ holds for all } z \text{ such that } |z|>\rho. \tag{6}$$

We now consider the function

$$g(z)=\frac{1}{f(z)-a}.$$

From (6) we see that $|g(z)|<\frac{1}{\varepsilon}$ for all z such that $|z|>\rho$. Within the circle K of radius ρ and center at 0, the function $g(z)$ may well have poles. In fact, it has a pole at every point where $f(z)=a$. The function $f(z)-a$ can have only a finite number of zeros inside the circle K. (If there were an infinite number, they would have a limit point, and the function $f(z)$ would be identically equal to a.) Let the zeros of $f(z)$ within K be at $\alpha_1, \alpha_2, ..., \alpha_m$. The function $g(z)$ has a pole at α_j, and so its Laurent expansion about the point α_j contains negative powers of the variable $z-\alpha_j$. Let $\varphi_j(z)$ be the sum of all of these negative powers, multiplied by the appropriate coefficients. Plainly we have $\lim \varphi_j(z)=0$ as $|z|\to\infty$. Let us now define

$$h(z)=g(z)-\varphi_1(z)-\varphi_2(z)-\cdots-\varphi_m(z).$$

The function $h(z)$ has no poles within the circle K and is certainly bounded on this circle and throughout its interior. The function $h(z)$ is also bounded throughout the exterior of the circle K. That is, $h(z)$ is a bounded entire function. By what we just proved, $h(z)$ is a constant. This implies that

$$\frac{1}{f(z)-a}=c_1+\varphi_1(z)+\varphi_2(z)+\cdots+\varphi_m(z),$$

the functions $\varphi_j(z)$ being polynomials in $\frac{1}{z-\alpha_j}$.

It follows that $f(z)$ itself has the form

$$f(z)=\psi(z),$$

where $\psi(z)$ is a rational function (that is, the quotient of two polynomials). The function $f(z)$ has no poles, being entire, and so the polynomial that is the denominator of $\psi(z)$ has no zeros except those that are cancelled out by zeros of the numerator. That is, $\psi(z)$ is simply a polynomial. We have already excluded this case. Thus our proof is complete.

We can also find a sequence of complex numbers satisfying (5) for which

$$\lim_{n \to \infty} |f(z_n)| = \infty.$$

If we deny this possibility, we see that $f(z)$ is bounded outside of a certain circle with center 0, and so $f(z)$ is bounded everywhere and therefore is a constant.

The Behavior of a Function Near an Isolated Singular Point. Let $F(z)$ be a function with an isolated singular point z_0. We use the results of §33 to write $F(z)$ in the form

$$F(z) = F_-(z) + F_+(z). \tag{7}$$

We carry out this decomposition somewhat differently from the way we carried it out in §33. That is, we take the constant term a_0 with the function $F_-(z)$ and not with the function $F_+(z)$. Thus the function $F_+(z)$, which has a Taylor series in positive powers of $z - z_0$, has the property that $F_+(z_0) = 0$. The series for $F_-(z)$,

$$F_-(z) = a_0 + a_{-1}(z - z_0)^{-1} + \cdots + a_{-n}(z - z_0)^{-n} + \cdots,$$

converges for all $z \neq z_0$, as we proved in §33. That is to say, the function

$$f(u) = a_0 + a_{-1}u + \cdots + a_{-n}u^n + \cdots$$

is an entire function. If the point z_0 is a pole of the function $F(z)$, then $f(u)$ is a polynomial. If z_0 is an essential singularity of the function $F(z)$, then the power series expansion of $f(u)$ contains an infinite number of nonzero coefficients. Now let a be an arbitrary complex number. From our earlier results ((4) and (5)) we can find a sequence of complex numbers

$$u_1, u_2, \ldots, u_n, \ldots$$

such that

$$\lim_{n \to \infty} |u_n| = \infty \tag{8}$$

and

$$\lim_{n \to \infty} f(u_n) = a. \tag{9}$$

We now write $z_n = z_0 + \dfrac{1}{u_n}$. From (8) we see that $\lim_{n \to \infty} z_n = z_0$, and from (9) that

$$\lim_{n \to \infty} F_-(z_n) = a.$$

Since $F_+(z_0)=0$, we have $\lim\limits_{n\to\infty} F_+(z_n)=0$. We thus have $\lim\limits_{n\to\infty} F(z_n)=a$. This phenomenon depends upon the hypothesis that z_0 be an essential singularity of the function $F(z)$. If z_0 is a pole of $F(z)$, we have $\lim\limits_{z\to z_0} |F(z)|=\infty$. Thus the behavior of a function near a pole is quite different from its behavior near an essential singularity.

Index

I. P. Cornfeld, S. V. Fomin, Y. G. Sinai

Ergodic Theory

Translated from the Russian
by A. B. Sossinskii

1981. X, 486 pages
(Grundlehren der mathematischen
Wissenschaften, Band 245)
ISBN 3-540-90580-4

Contents: Ergodicity and Mixing. Examples of Dynamic Systems. – Basic Constructions of Ergodic Theory. – Spectral Theory of Dynamical Systems. – Approximation Theory of Dynamical Systems by Periodic Dynamical Systems and Some of its Applications. – Appendices 1–4. – Bibliographical Notes. – Bibliography. – Index.

This book is a thorough exposition of the basic concepts and theorems from "abstract" ergodic theory including results on entropy, spectrum problems, and properties of approximation of dynamical systems by periodic dynamical systems. It includes many examples which familiarize the reader with theoretical notions and indicate how to use them in dealing with concrete examples of dynamical systems in various fields.

Springer-Verlag
Berlin
Heidelberg
New York
Tokyo

V. I. Arnold

Geometrical Methods in the Theory of Ordinary Differential Equations

Translated from the Russian by J. Szücs
English translation edited by M. Levi

1983. 153 figures. XI, 334 pages
(Grundlehren der mathematischen
Wissenschaften, Band 250)
ISBN 3-540-90681-9

Contents: Special Equations. – First-Order Partial Differential Equations. – Structural Stability. – Perturbation Theory. – Normal Forms. – Local Bifurcation Theory. – Samples of Examination Problems.

Written by one of the world's most famous analysts, this book develops a series of basic ideas and methods for the investigation of ordinary differential equations and their applications in the natural sciences. One pervading feature is the use of elementary methods of integration from the viewpoint of general mathematical concepts such as resolutions of singularities, Lie groups of symmetries, and Newton diagrams. At the center of the investigation is the qualitative theory of differential equations (structural stability, Anosov systems), asymptotic methods (averaging, adiabatic invariants), the analytic methods of the local theory in the neighborhood of a singular point or a periodic solution (Poincaré normal forms), and the theory of bifurcations of phase portraits in the variation of parameters. First order partial differential equations are investigated with the help of the geometry of a contact structure.

Arnold's book is addressed to a wide group of mathematicians applying differential equations in physics, the natural sciences and engineering.

Springer-Verlag
Berlin
Heidelberg
New York
Tokyo